Pituitary Today II: New Molecular, Physiological and Clinical Aspects

Frontiers of Hormone Research

Vol. 38

Series Editor

Ashley B. Grossman   London

Pituitary Today II

New Molecular, Physiological and Clinical Aspects

Volume Editors

Eduardo Arzt Buenos Aires
Marcello Bronstein Sao Paulo
Mirtha Guitelman Buenos Aires

39 figures, 2 in color, and 14 tables, 2010

Basel · Freiburg · Paris · London · New York · Bangalore ·
Bangkok · Shanghai · Singapore · Tokyo · Sydney

Eduardo Arzt, PhD
Laboratorio de Fisiología y Biología Molecular
Departamento de Fisiología y Biología y Celular
FCEN – Universidad de Buenos Aires
Ciudad Universiaria. Pabellon II
Buenos Aires, Argentina

Marcello Bronstein, MD
Neuroendocrine Unit,
Division of Endocrinology and Metabolism
Hospital das Clinicas
University of Sao Paulo Medical School
Sao Paulo, Brazil

Mirtha Guitelman, MD
Endocrinology Division
Durand Hospital, and Neuroscience's Institute
University of Buenos Aires (INUBA)
Buenos Aires, Argentina

Library of Congress Cataloging-in-Publication Data

Pituitary today II : new molecular, physiological and clinical aspects / volume editors Eduardo Arzt, Marcello Bronstein, Mirtha Guitelman.
     p. ; cm. -- (Frontiers of hormone research, ISSN 0301-3073 ; vol. 38)
  Other title: Pituitary today 2
  Other title: Pituitary today two
  Includes bibliographical references and index.
  ISBN 978-3-8055-9444-8 (hard cover : alk. paper)
  1. Pituitary gland--Diseases. 2. Pituitary gland--Cancer. 3. Pituitary gland--Molecular aspects. I. Arzt, Eduardo. II. Bronstein, Marcello D., 1945- III. Guitelman, Mirtha. IV. Title: Pituitary today 2. V. Title: Pituitary today two. VI. Series: Frontiers of hormone research, v. 38. 0301-3073 ;
  [DNLM: 1. Pituitary Diseases--physiopathology. 2. Pituitary Gland--physiopathology. W1 FR946F v.38 2010 / WK 500 P6929 2010]
  RC658.P577 2010
  616.4'7--dc22

2010006882

Bibliographic Indices. This publication is listed in bibliographic services, including Current Contents® and Index Medicus.

Disclaimer. The statements, opinions and data contained in this publication are solely those of the individual authors and contributors and not of the publisher and the editor(s). The appearance of advertisements in the book is not a warranty, endorsement, or approval of the products or services advertised or of their effectiveness, quality or safety. The publisher and the editor(s) disclaim responsibility for any injury to persons or property resulting from any ideas, methods, instructions or products referred to in the content or advertisements.

Drug Dosage. The authors and the publisher have exerted every effort to ensure that drug selection and dosage set forth in this text are in accord with current recommendations and practice at the time of publication. However, in view of ongoing research, changes in government regulations, and the constant flow of information relating to drug therapy and drug reactions, the reader is urged to check the package insert for each drug for any change in indications and dosage and for added warnings and precautions. This is particularly important when the recommended agent is a new and/or infrequently employed drug.

All rights reserved. No part of this publication may be translated into other languages, reproduced or utilized in any form or by any means electronic or mechanical, including photocopying, recording, microcopying, or by any information storage and retrieval system, without permission in writing from the publisher.

© Copyright 2010 by S. Karger AG, P.O. Box, CH–4009 Basel (Switzerland)
www.karger.com
Printed in Switzerland on acid-free and non-aging paper (ISO 9706) by Reinhardt Druck, Basel
ISSN 0301–3073
ISBN 978–3–8055–9444–8
e-ISBN 978–3–8055–9445–5

Contents

VII Foreword
Grossman, A.B. (London)

IX Preface
Guitelman, M. (Buenos Aires); Bronstein, M.D. (Sao Paulo); Arzt, E. (Buenos Aires)

1 Cytokines and Genes in Pituitary Tumorigenesis: RSUME Role in Cell Biology
Fuertes, M.; Gerez, J.; Haedo, M.; Giacomini, D. (Buenos Aires); Páez-Pereda, M.; Labeur, M.; Stalla, G.K. (Munich); Arzt, E. (Buenos Aires)

7 Molecular Mechanisms of Pituitary Adenoma Senescence
Chesnokova, V.; Zonis, S.; Ben-Shlomo, A.; Wawrowsky, K.; Melmed, S. (Los Angeles, Calif.)

15 Stem Cells, Differentiation and Cell Cycle Control in Pituitary
Drouin, J.; Bilodeau, S.; Roussel-Gervais, A. (Montréal, QC)

25 Role of Estrogens in Anterior Pituitary Gland Remodeling during the Estrous Cycle
Zárate, S.; Zaldivar, V.; Jaita, G.; Magri, L.; Radl, D.; Pisera, D.; Seilicovich, A. (Buenos Aires)

32 Hyperprolactinemia following Chronic Alcohol Administration
Sarkar, D.K. (New Brunswick, N.J.)

42 Experience from the Argentine Pegvisomant Observational Study: Preliminary Data
García Basavilbaso, N.; Guitelman, M.; Nagelberg, A.; Stalldecker, G.; Carabelli, A.; Bruno, O.; Danilowitz, K.; Manavela, M.; Mallea Gil, S.; Ballarino, C.; Guelman, R.; Katz, D.; Fidalgo, S.; Leal, R.; Fideleff, H.; Servidio, M. (Buenos Aires); Bruera, D. (Córdoba); Librandi, F.; Chervin, A.; Vitale, M.; Basso, A. (Buenos Aires)

50 Gender Differences in Macroprolactinomas: Study of Clinical Features, Outcome of Patients and Ki-67 Expression in Tumor Tissue
Fainstein Day, P.; Glerean, M.; Lovazzano, S.; Pietrani, M.; Christiansen, S.; Balzaretti, M.; Kozak, A.; Carrizo, A. (Buenos Aires)

59 Neurotransmitter Modulation of the GHRH-GH Axis
García-Tornadu, I.; Risso, G.; Perez-Millan, M.I.; Noain, D.; Diaz-Torga, G. (Buenos Aires); Low, M.J. (Portland, Oreg.); Rubinstein, M.; Becu-Villalobos, D. (Buenos Aires)

70 iASPP: A Novel Protein Involved in Pituitary Tumorigenesis?
Pinto, E.M.; Musolino, N.R.C.; Cescato, V.A.S.; Soares, I.C.; Wakamatsu, A.; de Oliveira, E.; Salgado, L.R.; Bronstein, M.D. (São Paulo)

77 Familial Isolated Pituitary Adenoma: Evidence for Genetic Heterogeneity
Toledo, R.A.; Lourenço, Jr., D.M.; Toledo, S.P.A. (São Paulo)

87 Serum Levels of 20K-hGH and 22K-hGH Isoforms in Acromegalic Patients
Lima, G.A.B.; Gadelha, M.R. (Rio de Janeiro); Strasburger, C.J.; Wu, Z. (Berlin)

94 Pituitary Carcinomas
Colao, A. (Naples); Ochoa, A.S. (Granada); Auriemma, R.S.; Faggiano, A.; Pivonello, R.; Lombardi, G. (Naples)

109 Modern Imaging of Pituitary Adenomas
Buchfelder, M.; Schlaffer, S.-M. (Erlangen)

121 Pathogenesis of Familial Acromegaly
Gadelha, M.R. (Rio de Janeiro); Frohman, L.A. (Chicago, Ill.)

127 Functional Role of the RET Dependence Receptor, GFRa Co-Receptors and Ligands in the Pituitary
Garcia-Lavandeira, M.; Diaz-Rodriguez, E.; Garcia-Rendueles, M.E.R.; Rodrigues, J.S.; Perez-Romero, S.; Bravo, S.B.; Alvarez, C.V. (Santiago de Compostela)

139 Testing Growth Hormone Deficiency in Adults
Gabellieri, E.; Chiovato, L. (Pavia); Lage, M.; Castro, A.I.; Casanueva, F.F. (Santiago de Compostela)

145 Serum Insulin-Like Growth Factor-1 Measurement in the Diagnosis and Follow-up of Patients with Acromegaly: Preliminary Data
Guitelman, M.; Radczuk, G.; Basavilbaso, N.G.; Oneto, A.; Basso, A. (Buenos Aires)

152 Diagnosis of Cure in Cushing's Syndrome: Lessons from Long-Term Follow-Up
Barahona, M.J.; Resmini, E.; Sucunza, N.; Webb, S.M. (Barcelona)

158 Novel Medical Therapies for Pituitary Tumors
Theodoropoulou, M.; Labeur, M.; Paez Pereda, M. (Munich); Haedo, M.; Perone, M.J. (Buenos Aires); Renner, U. (Munich); Arzt, E. (Buenos Aires); Stalla, G.K. (Munich)

165 Medical Therapy of Cushing's Disease: Where Are We Now?
Alexandraki, K.I.; Grossman, A.B. (London)

174 Optimizing Acromegaly Treatment
Bronstein, M.D. (Sao Paulo, SP)

184 Vasoinhibins and the Pituitary Gland
Méndez, I. (Querétaro, Qro/México, D.F.); Vega, C.; Zamorano, M.; Moreno-Carranza, B.; Martínez de la Escalera, G.; Clapp, C. (Querétaro, Qro)

190 Multiple Sources of Information for the Hypothalamus
van der Lelij, A.J. (Rotterdam)

196 New Insights in Ghrelin Orexigenic Effect
Diéguez, C. (Santiago de Compostela); da Boit, K.; Novelle, M.G.; Martínez de Morentin, P.B.; Nogueiras, R. (Barcelona); López, M. (Santiago de Compostela)

206 Ghrelin and Anterior Pituitary Function
Lanfranco, F.; Motta, G.; Baldi, M.; Gasco, V.; Grottoli, S.; Benso, A.; Broglio, F.; Ghigo, E. (Turin)

212 Author Index

214 Subject Index

Foreword

Some 2 years ago, at the end of the International Society of Endocrinology meeting in Rio de Janeiro, a group of us were transferred by a rather intermittently operating coach down the coast to the stunning resort of Angra dos Reis. There, for 2 days in November 2008, those who were fortunate to make this trip listened to and discussed the most recent and exciting advances in pituitary research, both basic and clinical. All the lectures at this meeting organized by Marcello Bronstein were pitched at the highest level, and we were able to have the intellectual joy of discussing our favourite subject within the confines of a most beautiful environment. Eduardo Arzt and Mirthe Guitelman are to be congratulated on bringing together such an inspiring group of lecturers, and we can now share much of their experience in this volume. Of course, you will miss out on the sun and beauty of Angra dos Reis, but I trust the intellectual excitement of reading about contemporary pituitary research written by from those deeply involved in the subject will be at least as valuable.

Ashley Grossman, London

Preface

On November 12–14, 2008, in Angra dos Reis, almost 200 professionals in medicine (endocrinologists, neurosurgeons), basic researchers, and clinical biochemists were hosted at the second edition of the scientific meeting 'Pituitary Today II: Molecular, Physiological and Clinical Aspects', where topics related to the pituitary gland have been discussed. Specialists from Latin America (among them from Argentina, Brazil, Mexico, Peru and Venezuela) gathered at this event, together with speakers from Canada, USA, Germany, England, Spain, Holland, Italy and Australia who attended the event as foreign guests.

Different aspects of pituitary physiopathology were covered in several symposia. Many of these lectures are presented in this book:

Genetics of pituitary tumors (Arzt, Melmed)

Pituitary cell development and organization in adult pituitary (Drouin, Seilicovich, Sarkar)

Research topics (Basavilbaso, Fainstein Day, Becu, Pinto, Toledo, Lima)

Pathogenesis and diagnostics of pituitary tumors (Colao, Buchfelder, Gadelha, Alvarez)

Diagnostics of pituitary disease: state of the art (Casanueva, Guitelman, Webb)

Therapeutic updates (Stalla, Grossman, Bronstein)

Hypothalamic control and pituitary hormones (Clapp, van der Lely, Dieguez, Gigho).

The meeting took place in a special environment where a scientific, cultural, and social exchange was predominant. We would like to mention that a group of young scientific fellows, both physicians and basic researchers, was also present and presented their results enriching the debates.

As coordinators of this event, we would like to express our thanks and recognition to Pfizer Laboratories who made it possible to hold the symposium 'Pituitary Today II' in Latin America; we also would like to mention the excellence of the staff whose working capacity enabled this meeting to reach an outstanding international level. We, as volume editors, thank all the contributors to this book in the Karger series *Frontiers in Hormone Research*.

Mirtha Guitelman, Buenos Aires
Marcello D. Bronstein, Sao Paulo
Eduardo Arzt, Buenos Aires

Cytokines and Genes in Pituitary Tumorigenesis: RSUME Role in Cell Biology

Mariana Fuertes[a] · Juan Gerez[a] · Mariana Haedo[a] · Damiana Giacomini[a] · Marcelo Páez-Pereda[b] · Marta Labeur[c] · Günter K. Stalla[c] · Eduardo Arzt[a]

[a]Laboratorio de Fisiología y Biología Molecular, Departamento de Fisiología y Biología Molecular y Celular, FCEN, Universidad de Buenos Aires, Buenos Aires, Argentina; [b]Affectis Pharmaceuticals, Munich, and [c]Max Planck Institute of Psychiatry, Munich, Germany

Abstract

Cytokines of the IL-6 or gp130 family regulate many cellular responses and play regulatory roles in numerous tissues, and are placed as auto-paracrine regulators of pituitary function acting in normal and tumoral anterior pituitary cells. Especially, IL-6 has a regulatory role in the hormone secretion and growth of the anterior pituitary and is involved in adenoma pathogenesis. Recently, IL-6 has been shown to mediate oncogene-induced senescence (OIS). IL-6 might participate in such a process in adenomas pituitary as well. From pituitary tumoral gp130 overexpressing cells, an unknown protein, RSUME, has been cloned. RSUME is induced by hypoxia in pituitary tumors and regulate pathways involved in angiogenic and tumorigenic processes (NF-kB/IkB and HIF-1α pathways). Thus, it could have an important role in the development of the pituitary tumors. Copyright © 2010 S. Karger AG, Basel

The gp130 Cytokines and Their Role in Pituitary

Cytokines exhibit functional pleiotrophy and redundancy, a particular cytokine may have a wide variety of biologic functions on various tissues and cells, and several different cytokines exert similar and overlapping functions on a certain cell. Cytokines that belong to the gp130 cytokine family (or the IL-6 cytokine family) have been shown to be synthesized in numerous organ systems such as hematopoietic tissues, reproductive tissues, thymus, heart, liver, pituitary, and the nervous system and are found to play roles in the regulation of cell differentiation, proliferation, cell survival, hormone secretion and inflammatory response [1]. These cytokines have also been shown to be involved in the development and growth of different tumors [2].

E.E.A. is Member of the IFYBINE-Argentine National Research Council (CONICET).

The IL-6 cytokine family is composed by IL-6, leukemia inhibitory factor (LIF), IL-11, ciliary neurotrophic factor (CNTF), oncostatin M, cardiotrophin-1 [3], neuropoietin [4] and cardiotrophin-like cytokine, also known as stimulating neurotrophin-1/B cell-stimulating factor-3 [5]. The binding of these cytokines to their specific receptors trigger the association of their alpha subunits with the membrane glycoprotein gp130, which functions as an initial cellular signal transducer [3]. Thus, this cytokine group is also named the gp130 cytokine family.

Considering that gp130 is ubiquitously expressed, the time and place at which gp130 functions in vivo appears to be determined by spatially and chronologically regulated expression of specific cytokine-binding receptor chains or cytokines themselves [2, 6, 7].

In the pituitary gland, many studies have demonstrated not only the expression of gp130 mRNA [8] but also the expression of specific receptors for IL-6 [9], LIF [10], IL-11 [11, 12], and CNTF [12]. Also, the synthesis of gp130 cytokines in different types of pituitary cells, such as LIF in developing human fetal pituitary and in normal and adenomatous human adult tissue [13], IL-11 in pituitary folliculostellate (FS), lactosomatotrophic and corticotrophic cells [11, 12] and CNTF in FS and lactosomatotrophic cells [12], has been shown. Particularly, several groups have demonstrated IL-6 production localized in FS cells of normal pituitary tissue [14], and in the tumor cells of pituitary adenomas [6, 15–19], and the presence of IL-6 mRNA in anterior pituitary cells [14, 15]. In addition, the JAK/STAT/SOCS-3 pathway, involved in the signal transduction of gp130 cytokines [3, 20], has also been described in pituitary cells [11, 21]. Therefore, the gp130 cytokines should be considered as auto-paracrine regulators of pituitary function in physiological and pathophysiological conditions, mainly in hormone secretion, cell growth, maintenance of homeostasis and tumorigenesis.

For example, the role of the cytokine transducer gp130 during the tumorigenic process in pituitary was demonstrated in a study where stable rat lactosomatotropic GH3 clones expressing different gp130 levels were generated and injected in athymic nude mice [22]. In contrast to mice injected with cells overexpressing gp130, those injected with cells expressing low gp130 levels showed a severely impaired in vivo tumor development. Furthermore, this clones with low gp130 levels showed reduced proliferation and hormone secretion (GH and prolactin) in response to gp130 cytokines. In contrast, the overexpression of gp130 did not significantly modify the cellular behavior, indicating that pituitary gp130 endogenous levels fulfill for a normal functional cellular response [22].

In addition, reduced levels of gp130 protein in MtT/S (pituitary somatotrophic cell line) cells stably transfected with gp130 antisense cDNA blocked cell growth and hormone secretion stimulated by CNTF and led to severely impaired in vivo tumor development in athymic nude mice [23]. These data, together, provide in vivo evidence supporting a link between gp130 and pituitary abnormal growth, and between gp130 and hormone secretion.

IL-6 Action in Pituitary Adenoma Growth

As stated above, the IL-6 cytokine and its receptor are produced in the pituitary gland, and regulate synthesis of anterior pituitary hormones, consistent with a paracrine or autocrine model. IL-6 directly regulates pituitary cell growth [6, 16, 17, 24]. In the normal pituitary, both paracrine and autocrine-derived IL-6 [6, 16] inhibit pituitary trophic growth. In contrast, in several tumor types (ACTH, PRL-secreting, GH-secreting and nonfunctioning adenomas), IL-6 exhibited either inhibitory or stimulatory effects not associated with tumor type or size [25].

Recently, it was reported that oncogene-induced senescence (OIS) is mediated by IL-6, activated in response to oncogenic stress [26]. This cytokine is required for both induction and maintenance of OIS, and acts in a cell-autonomous fashion to enable OIS. Experimental evidence suggests that IL-6 acts in an autocrine manner to regulate OIS as this signaling cascade is blocked by siIL-6 mRNA and requires an intact IL-6R. Thus, a cell-autonomous pool of IL-6 produced by senescent cells and acting in an autocrine and paracrine fashion mediates OIS [26].

Such a response to oncogenic stress, which restrains proliferation but allows the cell to remain viable and perform its physiological function, may be acting in the pituitary gland and thus favor vital functioning for homeostasis control. Pituitary adenoma senescence has been recently described [27, 28].

Given its role in pituitary growth, it is tempting to postulate the involvement of endogenous IL-6 in development of pituitary adenoma OIS, which, may explain the benign nature of these abundant tumors.

Cloning of Genes in Pituitary by mRNA Differential Display

Higher organisms contain many different genes, of which only a small fraction, perhaps 15%, are expressed in any individual cell. It is the choice of which genes are expressed that determines all life processes – development and differentiation, homeostasis, response to insults, cell cycle regulation, aging, and even programmed cell death. Altered gene expression lies at the heart of the regulatory mechanisms that control cell biology. Comparisons of gene expression in different cell types provide the underlying information we need to analyze the biological processes that control our lives [29].

mRNA differential display is potentially a powerful method for identifying genes that are over- or underexpressed in one mammalian cell type or tissue relative to another in distinct situations such as a specific developmental stage, certain time points or after an in vitro treatment. In this technique, the general strategy is to amplify partial cDNA sequences from subsets of mRNAs by reverse transcription and the polymerase chain reaction (PCR) [29, 30].

This approach resulted effective in the pituitary. First, a comparison of the gene expression pattern between normal anterior pituitaries and prolactinomas using, as a model of this latter, female dopamine D2-receptor-deficient mice (D2R–/–), allowed the identification of a band present only in the normal pituitary tissue as noggin, a specific inhibitor of bone morphogenetic protein 4 (BMP-4) [31]. BMP4 was further showed to stimulate lactotrophic and inhibit corticotrophic cell growth, having an opposite effect in both cell types [31, 32].

Recently, this cloning technique also was used for compared one stable clone from the rat pituitary lactosomatotrophic tumor cell line GH3 overexpressing gp130, that showed increased tumorigenic and angiogenic potential when injected into nude mice [22], to a control GH3 clone stable for the empty vector. In a GH3 clone that overexpressed gp130 appeared a differential band corresponding to a previously uncharacterized mRNA that was called RSUME [33].

RSUME Characterization and Function

The unknown gene cloned from the pituitary cell line, RSUME, has two splicing variants in humans of 195 or 267 amino acids, coding for a small RWD domain-containing protein. Its name is based on its structure and function (RSUME for RWD-domain-containing sumoylation enhancer). This new protein is highly conserved in higher vertebrates, does not seem to belong to any previously known protein family. Also has been demonstrated by confocal microscopy that RSUME protein is located in the cytoplasm and nucleus, despite lacking a nuclear localization signal [33].

RSUME is expressed in various tissues, but the higher expression levels were found in cerebellum, pituitary, heart, kidney, liver, stomach, pancreas, prostate, and spleen. It was upregulated by cellular stress stimuli such as hypoxia and heat shock. Particularly, in pituitary tumors RSUME expression was increased by hypoxia [33].

RSUME enhances protein sumoylation through a direct interaction with Ubc9, a SUMO conjugase. Moreover, the correct folding of its RWD domain is essential for RSUME activity. RSUME has its strongest effect in the step of forming the Ubc9-SUMO-1 thioester, but also acted in the transference of SUMO-1 from the thioester to a substrate [33].

RSUME increased sumoylation and protein stability of IkB, resulting in inhibition of NF-kB transcriptional activity and, consequently, suppressing the inflammatory process mediated by NF-kB targets such as interleukin-8 (IL-8) and cyclooxigenase-2 (Cox-2). Furthermore, RSUME plays a role in the cell's response to hypoxic stress because it increased hypoxia inducible factor-1α (HIF-1α) protein levels and enhanced it sumoylation and stability, thus regulating adaptive responses to changes in oxygen tension and angiogenesis in mammalian cells [33].

Given the fact that the actions of RSUME in sumoylation have functional implications on the NF-kB/IkB pathway and on the regulation of HIF-1α, both involved in angiogenic and tumorigenic processes, its expression and induction by hypoxia in pituitary tumors and its cloning from gp130 overexpressing pituitary tumor cells, RSUME mechanism of participation in adenoma pathogenesis is under study.

Acknowledgements

This work was supported by grants from the University of Buenos Aires, the Argentine National Research Council (CONICET) and Agencia Nacional de Promoción Científica y Tecnológica, Argentina.

References

1 Heinrich PC, Behrmann I, Muller-Newen G, Schaper F, Graeve L: Interleukin-6-type cytokine signalling through the gp130/Jak/STAT pathway. Biochem J 1998;334:297–314.
2 Carbia-Nagashima A, Arzt E: Intracellular proteins and mechanisms involved in the control of gp130/JAK/STAT cytokine signaling. IUBMB Life 2004;56:83–88.
3 Kishimoto T, Akira S, Narazaki M, Taga T: Interleukin-6 family of cytokines and gp130. Blood 1995;86:1243–1254.
4 Derouet D, Rousseau F, Alfonsi F, Froger J, Hermann J, Barbier F, Perret D, Diveu C, Guillet C, Preisser L, Dumont A, Barbado M, Morel A, de Lapeyriere O, Gascan H, Chevalier S: Neuropoietin, a new IL-6-related cytokine signaling through the ciliary neurotrophic factor receptor. PNAS 2004;101:4827–4832.
5 Senaldi G, Varnum BC, Sarmiento U, Starnes C, Lile J, Scully S, Guo J, Elliott G, McNinch J, Shaklee CL, Freeman D, Manu F, Simonet WS, Boone T, Chang MS: Novel neurotrophin-1/B cell-stimulating factor-3: a cytokine of the IL-6 family. Proc Natl Acad Sci USA 1999;96:11458–11463.
6 Arzt E, Páez-Pereda M, Perez-Castro C, Pagotto U, Renner U, Stalla GK: Pathophysiological role of the cytokine network in the anterior pituitary gland. Front Neuroendocrinol 1999;20:71–95.
7 Arzt E, Stalla GK: Cytokines: autocrine and paracrine roles in the anterior pituitary. Neuroimmunomodulation 1996;3:28–34.
8 Shimon I, Yan X, Ray D, Melmed S: Cytokine-dependent gp130 receptor subunit regulates human fetal pituitary adrenocorticotropin hormone and growth hormone secretion. J Clin Invest 1997;100:357–363.
9 Ohmichi M, Hirota K, Koike K, Kurachi H, Ohtsuka S, Matsuzaki N, Yamaguchi M, Miyake A, Tanizawa O: Binding sites for interleukin-6 in the anterior pituitary gland. Neuroendocrinology 1992;55:199–203.
10 Akita S, Webster J, Ren SG, Takino H, Said J, Zand O, Melmed S: Human and murine pituitary expression of leukemia inhibitory factor: novel intrapituitary regulation of adrenocorticotropin hormone synthesis and secretion. J Clin Invest 1995;95:1288–1298.
11 Auernhammer CJ, Melmed S: Interleukin-11 stimulates proopiomelanocortin gene expression and adrenocorticotropin secretion in corticotroph cells: evidence for a redundant cytokine network in the hypothalamo-pituitary-adrenal axis. Endocrinology 1999;140:1559–1566.
12 Perez-Castro C, Carbia-Nagashima A, Páez-Pereda M, Goldberg V, Chervin A, Largen P, Renner U, Stalla GK, Arzt E: The gp130 cytokines interleukin-11 and ciliary neurotropic factor regulate through specific receptors the function and growth of lactosomatotropic and folliculostellate pituitary cell lines. Endocrinology 2000;141:1746–1753.
13 Akita S, Conn PM, Melmed S: Leukemia inhibitory factor (LIF) induces acute adrenocorticotrophic hormone (ACTH) secretion in fetal rhesus macaque primates: a novel dynamic test of pituitary function. J Clin Endocrinol Metab 1996;81:4170–4178.
14 Vankelecom H, Carmeliet P, Van Damme J, Billiau A, Denef C: Production of interleukin-6 by folliculo-stellate cells of the anterior pituitary gland in a histiotypic cell aggregate culture system. Neuroendocrinology 1989;49:102–106.
15 Spangelo BL, Gorospe WC: Role of the cytokines in the neuroendocrine-immune system axis. Front Neuroendocrinol 1995;16:1–12.
16 Arzt E: gp130 cytokine signaling in the pituitary gland: a paradigm for cytokine-neuro-endocrine pathways. J Clin Invest 2001;108:1729–1733.
17 Arzt E, Buric R, Stelzer G, Stalla J, Sauer J, Renner U, Stalla GK: Interleukin involvement in anterior pituitary cell growth regulation: effects of IL-2 and IL-6. Endocrinology 1993;132:459–467.

18 Renner U, Gloddek J, Páez-Pereda M, Arzt E, Stalla GK: Regulation and role of intrapituitary IL-6 production by folliculostellate cells. Domestic Anim Endocrinol 1998;51:353–362.
19 Jones TH, Daniels M, James RA, Justice SK, McCorkle R, Price A, Kendall-Taylor P, Weetman AP: Production of bioactive and immunoreactive interleukin-6 (IL-6) and expression of IL-6 messenger ribonucleic acid by human pituitary adenomas. J Clin Endocrinol Metab 1994;78:180–187.
20 Levy DE, Darnell JE: Stats: transcriptional control and biological impact. Nat Rev Mol Cell Biol 2002;3:651–662.
21 Busquet C, Zatelli MC, Melmed S: Direct regulation of pituitary proopiomelanocortin by STAT3 provides a novel mechanism for immunoneuroendocrine interfacing. J Clin Invest 2000;106:1417–1425.
22 Perez-Castro C, Giacomini D, Carbia-Nagashima AC, Onofri C, Graciarena M, Kobayashi K, Páez-Pereda M, Renner U, Stalla GK, Arzt E: Reduced expression of the cytokine transducer gp130 inhibits hormone secretion, cell growth, and tumor development of pituitary lactosomatotrophic GH3 cells. Endocrinology 2003;144:693–700.
23 Graciarena M, Carbia-Nagashima A, Onofri C, Perez-Castro C, Giacomini D, Renner U, Stalla GK, Arzt E: Involvement of the gp130 cytokine transducer in MtT/S pituitary somatotroph tumour development in an autocrine-paracrine model. Eur J Endocrinol 2004;151:595–604.
24 Arzt E, Sauer J, Buric R, Stalla J, Renner U, Stalla GK: Characterization of Interleukin-2 (IL-2) receptor expression and action of IL-2 and IL-6 on normal anterior pituitary cell growth. Endocrine 1995;3:113–119.
25 Páez-Pereda M, Goldberg V, Chervín A, Carrizo G, Molina A, Andrada J, Sauer J, Renner U, Stalla GK, Arzt E: Interleukin-2 (IL-2) and IL-6 regulate c-fos proto-oncogene expression in human pituitary adenoma explants. Mol Cell Endocrinol 1996;124:33–42.
26 Kuilman T, Michaloglou C, Vredeveld LC, Douma S, van Doorn R, Desmet CJ, Aarden LA, Mooi WJ, Peeper DS: Oncogene-induced senescence relayed by an interleukin-dependent inflammatory network. Cell 2008;133:1019–1031.
27 Chesnokova V, Zonis S, Rubinek T, Yu R, Ben-Shlomo A, Kovacs K, Wawrowsky K, Melmed S: Senescence mediates pituitary hypoplasia and restrains pituitary tumor growth. Cancer Res 2007;67:10564–10572.
28 Chesnokova V, Zonis S, Kovacs K, Ben-Shlomo A, Wawrowsky K, Bannykh S, Melmed S: p21(Cip1) restrains pituitary tumor growth. Proc Natl Acad Sci USA 2008;105:17498–17503.
29 Liang P, Pardee AB: Differential display of eukaryotic messenger RNA by means of the polymerase chain reaction. Science 1992;257:967–971.
30 Sompayrac L, Jane S, Burn TC, Tenen DG, Danna KJ: Overcoming limitations of the mRNA differential display technique. Nucl Acids Res 1995;23:4738–4739.
31 Páez-Pereda M, Giacomini D, Refojo D, Nagashima AC, Hopfner U, Grubler Y, Chervin A, Goldberg V, Goya R, Hentges ST, Low MJ, Holsboer F, Stalla GK, Arzt E: Involvement of bone morphogenetic protein-4 in pituitary prolactinoma pathogenesis through a Smad/estrogen receptor crosstalk. Proc Natl Acad Sci USA 2003;100:1034–1039.
32 Giacomini D, Páez-Pereda M, Theodoropoulou M, Labeur M, Refojo D, Gerez J, Chervin A, Berner S, Losa M, Buchfelder M, Renner U, Stalla GK, Arzt E: Bone morphogenetic protein-4 inhibits corticotroph tumor cells: involvement in the retinoic acid inhibitory action. Endocrinology 2006;147:247–256.
33 Carbia-Nagashima A, Gerez J, Perez-Castro C, Páez-Pereda M, Silberstein S, Stalla GK, Holsboer F, Arzt E: RSUME, a small RWD-containing protein, enhances SUMO conjugation and stabilizes HIF-1α during hypoxia. Cell 2007;131:309–323.

Dr. Eduardo Arzt
Laboratorio de Fisiología y Biología Molecular, FCEN
Universidad de Buenos Aires, Ciudad Universitaria
1428 Buenos Aires (Argentina)
Tel. +54 11 4576 3368/86, Fax +54 11 4576 3321, E-Mail earzt@fbmc.fcen.uba.ar

Molecular Mechanisms of Pituitary Adenoma Senescence

Vera Chesnokova · Svetlana Zonis · Anat Ben-Shlomo · Kolja Wawrowsky · Shlomo Melmed

Department of Medicine, Cedars-Sinai Medical Center, Los Angeles, Calif., USA

Abstract

As commonly encountered pituitary adenomas are invariably benign, we examined protective pituitary proliferative mechanisms. Cellular senescence is characterized by a largely irreversible cell cycle arrest and constitutes a strong anti-proliferative response, which can be triggered by DNA damage, chromosomal instability and aneuploidy, loss of tumor suppressive signaling or oncogene activation. Cellular senescence may prevent cells from undergoing transformation in vitro. Recently, evidence has accumulated that in vivo senescence is an important protective mechanisms against cancer. In this review we highlight an intrinsic predisposition of pituitary tumors to exhibit senescence-associated molecular pathways and show prospective mechanisms underlying the benign nature of these commonly encountered tumors.

Copyright © 2010 S. Karger AG, Basel

Sporadic pituitary adenomas are present in 25% of autopsy specimens. Tumors may arise from any of five highly differentiated pituitary cell subtypes. Excess pituitary hormone secretion is usually associated with invariably benign monoclonal adenomas. The prevalence of pituitary adenomas increases with advancing age and both sexes are affected equally [1, 2]. PRL-producing adenomas are the most common type, while approximately 30% of all adenomas are not associated with hypersecretion, and most of these are immunopositive for LH-/FSH. ACTH- and GH-producing adenomas each account for 10–15% of all adenomas, and TSH-producing adenomas are extremely rare. Although pituitary chromosome instability is an early hallmark of pituitary adenomas development and growth, pituitary adenoma rarely progress to true carcinoma [3].

Several factors may account for the biological indolence of these neoplasms. Progression to a malignant phenotype requires multiple oncogenic mutations, which rarely occur. Indeed, with the exception of *gsp* oncogene in subset of somatotrophinomas [4], activating mutations of oncogenes, or mutations resulting in inactivation of tumor suppressor genes have not been found. Tumor growth requires adequate vascularization to supply oxygen to tumor cells. However, in contrast to other tumors, pituitary adenomas are less vascularized then surrounding normal tissues [5]. It is also plausible that intrinsic properties of highly differentiated and specialized pituitary cells limit their ability for uncontrolled proliferation.

Cellular Senescence

In contrast to germ cells, most somatic cells permanently stop dividing after a finite number of cell divisions in culture (Hayflick limit) [6], and enter a state termed cellular or replicative senescence. These cells are irreversibly arrested in G1 phase of the cell cycle and no longer divide, despite remaining viable and metabolically active for long periods of time. Cellular senescence, or proliferation arrest takes place in aging tissue due to telomere shortening, and also occurs in response to externals stressors, such as oxidative or genotoxic stress, DNA damage, aneuploidy or chromosomal instability [7, 8].

In addition, cellular senescence can be also triggered by other types of stress, including oncogene activation. Oncogenic RAS elicits stable proliferative arrest rather than oncogenic transformation in young diploid fibroblasts in vitro [9]. Overexpression of oncogene BRAF in cultured human melanocytes caused growth arrest, and BRAF induction was detected in vivo in benign human nevi, which eventually lose proliferative activity and their growth may remain arrested for decades. Cellular senescence is induced in premalignant tumors but disappear in more advanced malignant tumors [10–12]. Most tumors contain cells that appear to have bypassed this limit and evade senescence [13]. Therefore cellular senescence might represent an initial impediment against oncogenic stimulation and might play an important role in tumor suppression [14, 15].

Cellular Senescence Machinery

Rb is a tumor suppressor gene and acts as a G1/S checkpoint control. Cyclin-dependent kinases (Cdk) phosphorylate Rb to release E2F, enabling S phase and ptoliferation to progress. Cell-cycle progression is regulated by sequential activation and inactivation of protein kinase complexes including binding to Cdk inhibitors that govern the G1-to S transition. Cdk actions are suppressed by INK4/ARF-type inhibitors (p16, p18, p19) and Cip/Kip-type (p21, p27, p57) inhibitors [16]. These proteins arrest cell cycle progression, and act as growth suppressor genes.

Premature senescence manifests by activation of a set of tumor suppressor genes that are often inactivated in human cancer; products of the INK4A/ARF locus that encode proteins p16 and p19 with alternative reading frame, p53, or Rb [17–19].

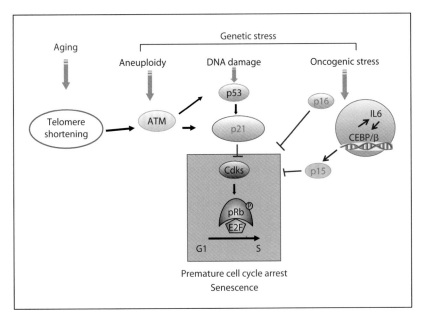

Fig. 1. Cartoon depicting multiple senescence pathways.

The Cdk inhibitor p21 which belongs to Cip/Kip family is one of the main p53 targets for induction of cell cycle arrest following DNA damage. Pathways that generate DNA damage response and p53 activation utilize p21 as a major mediator of cellular senescence to control Rb protein phosphorylation [20]. Studies have revealed a role of the inflammatory transcriptome in oncogene-induced senescence (OIS). This latter mechanism of induction is of particular interest because it suggests that not only intrinsic cellular factors but also extracellular and secreted proteins, including IGFBP7, TGFβ, IL-8 and IL-6 are autocrine mediators of oncogene-induced senescence [21]. Cells undergoing OIS communicate with the environment producing inflammatory cytokines, and the transcription factor C/EBPβ cooperates with IL-6 to amplify activation of an inflammatory network which maintains cellular senescence [21] (fig. 1).

Pituitary Tumor Transforming Gene

The pituitary tumor transforming gene (PTTG) protein was initially isolated from pituitary tumor cells [22], and behaves as a mammalian securin homolog facilitating sister chromatid separation during metaphase [23]. Both PTTG deletion and overexpression lead to dysregulated G2/M checkpoint surveillance, resulting in abnormal mitosis, aneuploidy and chromosomal instability [24, 25]. In contrast to restricted normal tissue expression, PTTG is abundantly expressed in pituitary, thyroid, breast esophageal, endometrial and colorectal tumors and correlates with tumor invasiveness [26]. Notably, *Pttg* was identified as a key signature gene associated with tumor metastases [27]. PTTG induces basic fibroblast growth factor (bFGF) [28, 29], activates angiogenesis, and activates c-myc and CCND3, enhancing cell proliferation [22, 30, 31]. *Pttg* exhibits oncogene properties as it causes cell transformation in vitro, and promotes tumor formation in nude mice [22]. Using the glycoprotein hormone α-subunit promoter, we generated transgenic mice with pituitary-driven *Pttg* overexpression [32, 33]. We documented clear magnetic resonance imaging and histological evidence for pituitary hyperplasia preceding adenoma formation. *Pttg* overexpression resulted in focal pituitary hyperplasia and LH, GH and TSH-secreting adenoma formation with respective trophic hormone hypersecretion [32].

Pituitary Tumor Senescence

Pttg deletion results in pituitary hypoplasia, low pituitary cell proliferation rates, and aneuploidy. Aneuploidy and chromosomal instability in *Pttg*-null pituitary glands triggers DNA damage signaling pathway activation including p53 accumulation and p21 induction [34, 35]. *Pttg* deletion rescued pituitary and thyroid tumor development in *Rb+/–* mice, which usually develop tumors with high penetrance. In the *Pttg*-null pituitary, intra-nuclear p21 was induced. High pituitary p21 levels in the absence of PTTG were associated with suppressed Cdk2 activity, Rb phosphorylation and cyclin A expression, all required for cell cycle progression. Continuing p21 induction restrains cell cycle progression and triggers irreversible pituitary cell proliferation arrest as evidenced by high levels of senescence-associated β-galactosidase (SA-β-gal) activity. However, *Pttg*-deficient pituitary senescence is not associated with telomere shortening, supporting the presence of premature pituitary cell senescence [35]. In contrast, deletion of p21 from *Rb+/–Pttg–/–* mice showed increased pituitary cell proliferation and decreased senescence as evidenced in the mouse embryonic fibroblasts (MEF) colony assay. p21 deletion completely restored pituitary tumor development in *Rb+/–Pttg–/–* mice to levels observed in *Rb+/–* animals, indicating that p21 and p53/p21 senescence pathways play a key role restraining tumor growth [36, 37].

Interestingly, intranuclear p21 selectively accumulates in *Pttg*-null pituitary GH-secreting cells. Human GH-producing pituitary adenomas over-express PTTG, and also exhibit aneupolidy and senescence evidenced by increased p21, kinase mutated in ataxia-telangiectasia (ATM), and SA-β-gal levels. In contrast, p21 was undetectable in the human pituitary carcinomas tested [36]. Thus, dysregulation of PTTG, as evidenced in both pituitary tumors and in PTTG-null pituitary glands, leads to aneuploidy, DNA damage and activation of the senescence pathway (fig. 2). These results suggest that aneuploid pituitary cell p21 may constrain pituitary tumor growth, thus providing a proposed mechanism for the very low observed incidence of pituitary carcinomas.

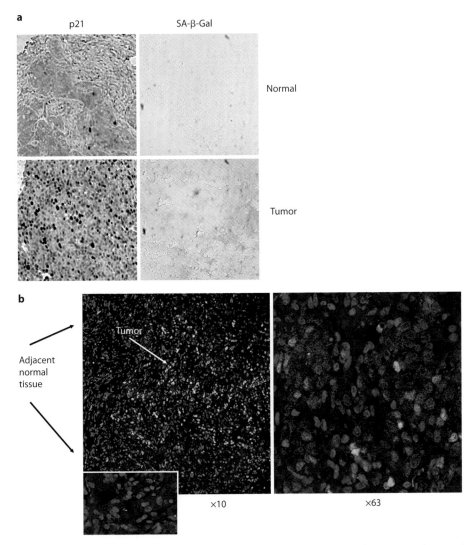

Fig. 2. a Immunohistochemistry of the same GH-secreting human adenoma sections stained for p21 (brown), and SA-β-gal activity (blue). **b** Confocal image of double fluorescence immunohistochemistry of p21 (green) and β-galactosidase (red) proteins co-expression in human pituitary adenomatous but not in normal adjacent tissue. Right panel: high-resolution (×63) image of the same slide.

Multiple senescence pathways might trigger irreversible pituitary cell proliferation arrest. Activated DNA damage pathways and p53/p21 senescence might be induced by securin properties of PTTG. On the other hand, PTTG behaves a protooncogene, and high levels of PTTG can trigger oncogene-induced senescence. Similarly, Lazzerini Denchi et al. [38] reported an activation of senescence, including induction of p16 and p19, in pituitary gland that were driven by forced expression of the growth promoting transcription factor E2F3. Oncogene-induced

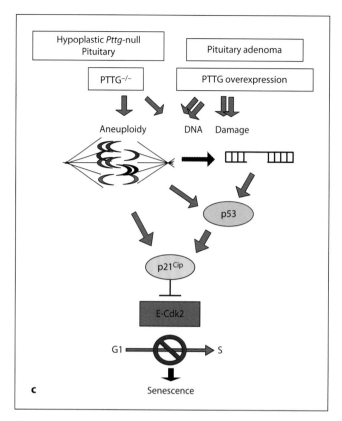

Fig. 2. c Proposed model for p21-induced senescence in the hypoplastic *Pttg*-null pituitary gland and PTTG-overexpressing pituitary adenomas [36].

senescence is enabled via p16/Rb-[39], or IL6-C/EBPβ senescence pathways [21, 39]. Members of the gp130 cytokine family including LIF, IL-6, IL-11 and CNTF control pituitary cell division and regulate synthesis of anterior pituitary hormones [40]. IL-6 directly regulates pituitary cell growth [41]. Although IL-6 stimulates DNA synthesis and the GH3 cell number, similar concentrations of IL-6 actually inhibit growth of normal pituitary cells. In several pituitary tumor types, IL-6 exhibits either inhibitory or stimulatory effects. Considering the modest adult pituitary cell proliferation, pituitary cell growth regulation by IL-6 underlies the role of cytokines as factors controlling pituitary cell division. The new finding of the role of IL-6 in OIS suggests possible involvement of IL-6 in the development of pituitary adenoma senescence [42].

Thus, pituitary adenomas constitute faithful in vivo model of senescence. Pituitary cells are one of the few epithelial cell types that do not readily undergo malignant transformation. Pituitary cells undergo premature senescence enabling escape from the proliferative pressure of oncogenes, hormones and transformation factors. Senescence is considered an important tumor protection barrier in several models

[10, 15, 20, 39, 43]. Understanding mechanisms underlying the ability of pituitary cells to escape aggressive growth and malignant transformation may provide important insights into cancer-restraining pathways and offer new opportunities for subcellular therapeutic approaches.

References

1 Asa SL, Ezzat S: The pathogenesis of pituitary tumors. Annu Rev Pathol 2009;4:97–126.
2 Lania AG, Mantovani G, Spada A: Mechanisms of disease: Mutations of G proteins and G-protein-coupled receptors in endocrine diseases. Nat Clin Pract Endocrinol Metab 2006;2:681–693.
3 Melmed S: Mechanisms for pituitary tumorigenesis: the plastic pituitary. J Clin Invest 2003;112:1603–1618.
4 Landis CA, Masters SB, Spada A, Pace AM, Bourne HR, Vallar L: GTPase inhibiting mutations activate the alpha chain of Gs and stimulate adenylyl cyclase in human pituitary tumours. Nature 1989;340:692–696.
5 Viacava P, Gasperi M, Acerbi G, et al: Microvascular density and vascular endothelial growth factor expression in normal pituitary tissue and pituitary adenomas. J Endocrinol Invest 2003;26:23–28.
6 Hayflick L, Perkins F, Stevenson RE: Human Diploid Cell Strains. Science 1964;143:976.
7 Sharpless NE, DePinho RA: Telomeres, stem cells, senescence, and cancer. J Clin Invest 2004;113:160–168.
8 Serrano M, Blasco MA: Putting the stress on senescence. Curr Opin Cell Biol 2001;13:748–753.
9 Serrano M, Lin AW, McCurrach ME, Beach D, Lowe SW: Oncogenic ras provokes premature cell senescence associated with accumulation of p53 and p16INK4a. Cell 1997;88:593–602.
10 Collado M, Gil J, Efeyan A, et al: Tumour biology: senescence in premalignant tumours. Nature 2005;436:642.
11 Braig M, Lee S, Loddenkemper C, et al: Oncogene-induced senescence as an initial barrier in lymphoma development. Nature 2005;436:660–665.
12 Michaloglou C, Vredeveld LC, Soengas MS, Denoyelle C, Kuilman T, van der Horst CM, Majoor DM, Shay JW, Mooi WJ, Peeper DS: BRAFE600-associated senescence-like cell cycle arrest of human naevi. Nature 2005;436:720–724.
13 Ohtani N, Mann DJ, Hara E: Cellular senescence: its role in tumor suppression and aging. Cancer Sci 2009;100:792–797.
14 Campisi J: Cellular senescence as a tumor-suppressor mechanism. Trends Cell Biol 2001;11:S27–S31.
15 Campisi J: Suppressing cancer: the importance of being senescent. Science 2005;309:886–887.
16 Sherr CJ, Roberts JM: CDK inhibitors: positive and negative regulators of G1-phase progression. Genes Dev 1999;13:1501–1512.
17 Schmitt CA, Fridman JS, Yang M, Lee S, Baranov E, Hoffman RM, Lowe SW: A senescence program controlled by p53 and p16INK4a contributes to the outcome of cancer therapy. Cell 2002;109:335–346.
18 Palmero I, Pantoja C, Serrano M: p19ARF links the tumour suppressor p53 to Ras. Nature 1998;395:125–126.
19 Kiyokawa H, Koff A: Roles of cyclin-dependent kinase inhibitors: lessons from knockout mice. Curr Top Microbiol Immunol 1998;227:105–120.
20 Mooi WJ, Peeper DS: Oncogene-induced cell senescence–halting on the road to cancer. N Engl J Med 2006;355:1037–1046.
21 Kuilman T, Peeper DS: Senescence-messaging secretome: SMS-ing cellular stress. Nat Rev Cancer 2009;9:81–94.
22 Pei L, Melmed S: Isolation and characterization of a pituitary tumor-transforming gene (PTTG). Mol Endocrinol 1997;11:433–441.
23 Zou H, McGarry TJ, Bernal T, Kirschner MW: Identification of a vertebrate sister-chromatid separation inhibitor involved in transformation and tumorigenesis. Science 1999;285:418–422.
24 Yu R, Heaney AP, Lu W, Chen J, Melmed S: Pituitary tumor transforming gene causes aneuploidy and p53-dependent and p53-independent apoptosis. J Biol Chem 2000;275:36502–36505.
25 Yu R, Lu W, Chen J, McCabe CJ, Melmed S: Overexpressed pituitary tumor-transforming gene causes aneuploidy in live human cells. Endocrinology 2003;144:4991–4998.
26 Vlotides G, Eigler T, Melmed S: Pituitary tumor-transforming gene: physiology and implications for tumorigenesis. Endocr Rev 2007;28:165–186.
27 Ramaswamy S, Ross KN, Lander ES, Golub TR: A molecular signature of metastasis in primary solid tumors. Nat Genet 2003;33:49–54.
28 Ishikawa H, Heaney AP, Yu R, Horwitz GA, Melmed S: Human pituitary tumor-transforming gene induces angiogenesis. J Clin Endocrinol Metab 2001;86:867–874.

29 Heaney AP, Horwitz GA, Wang Z, Singson R, Melmed S: Early involvement of estrogen-induced pituitary tumor transforming gene and fibroblast growth factor expression in prolactinoma pathogenesis. Nat Med 1999;5:1317–1321.

30 Hamid T, Kakar SS: PTTG/securin activates expression of p53 and modulates its function. Mol Cancer 2004;3:18.

31 McCabe CJ, Khaira JS, Boelaert K, Heaney AP, Tannahill LA, Hussain S, Mitchell R, Olliff J, Sheppard MC, Franklyn JA, et al: Expression of pituitary tumour transforming gene (PTTG) and fibroblast growth factor-2 (FGF-2) in human pituitary adenomas: relationships to clinical tumour behaviour. Clin Endocrinol (Oxf) 2003;58:141–150.

32 Abbud RA, Takumi I, Barker EM, Ren SG, Chen DY, Wawrowsky K, Melmed S: Early multipotential pituitary focal hyperplasia in the alpha-subunit of glycoprotein hormone-driven pituitary tumor-transforming gene transgenic mice. Mol Endocrinol 2005;19:1383–1391.

33 Donangelo I, Melmed S: Implication of pituitary tropic status on tumor development. Front Horm Res 2006;35:1–8.

34 Chesnokova V, Kovacs K, Castro AV, Zonis S, Melmed S: Pituitary hypoplasia in Pttg–/– mice is protective for Rb+/– pituitary tumorigenesis. Mol Endocrinol 2005;19:2371–2379.

35 Chesnokova V, Zonis S, Rubinek T, Yu R, Ben-Shlomo A, Kovacs K, Wawrowsky K, Melmed S: Senescence mediates pituitary hypoplasia and restrains pituitary tumor growth. Cancer Res 2007;67:10564–10572.

36 Chesnokova V, Zonis S, Kovacs K, Ben-Shlomo A, Wawrowsky K, Bannykh S, Melmed S: p21(Cip1) restrains pituitary tumor growth. Proc Natl Acad Sci USA 2008;105:17498–17503.

37 Chesnokova V, Melmed S: Pituitary tumour-transforming gene (PTTG) and pituitary senescence. Horm Res 2009;71(suppl 2):82–87.

38 Lazzerini Denchi E, Attwooll C, Pasini D, Helin K: Deregulated E2F activity induces hyperplasia and senescence-like features in the mouse pituitary gland. Mol Cell Biol 2005;25:2660–2672.

39 Collado M, Blasco MA, Serrano M: Cellular senescence in cancer and aging. Cell 2007;130:223–233.

40 Arzt E: gp130 cytokine signaling in the pituitary gland: a paradigm for cytokine-neuro-endocrine pathways. J Clin Invest 2001;108:1729–1733.

41 Arzt E, Pereda MP, Castro CP, Pagotto U, Renner U, Stalla GK: Pathophysiological role of the cytokine network in the anterior pituitary gland. Front Neuroendocrinol 1999;20:71–95.

42 Arzt E, Chesnokova V, Stalla GK, Melmed S: Pituitary adenoma growth: a model for cellular senescence and cytokine action. Cell Cycle 2009;8:677–678.

43 Cichowski K, Hahn WC: Unexpected pieces to the senescence puzzle. Cell 2008;133:958–961.

Shlomo Melmed, MD
Academic Affairs, Room 2015, Cedars-Sinai Medical Center
8700 Beverly Blvd.
Los Angeles, CA 90048 (USA)
Tel. +1 310 423 4691, Fax +1 310 423 0119, E-Mail melmed@csmc.edu

Stem Cells, Differentiation and Cell Cycle Control in Pituitary

Jacques Drouin · Steve Bilodeau · Audrey Roussel-Gervais

Institut de Recherches Cliniques de Montréal (IRCM), Montréal, QC, Canada

Abstract

As model of organogenesis, the pituitary gland is a relatively simple tissue; yet, we understand little of the mechanisms that determine organ size, cell number and allocation of cells to different lineages. While the discovery of cell-restricted transcription factors has led to significant insight into the mechanisms controlling differentiation and cell-specific gene expression, we still need to integrate these processes with control of organ development. The identification of pituitary stem cells has suggested mechanisms for maintenance of adult pituitary but these findings again highlight the crucial role of cell cycle control for determination of progenitor and differentiated cell numbers. We recently described the mechanisms for progenitor cell cycle exit in early pituitary development that critically depend on the cell cycle inhibitor p57^{Kip2}. It appears that cell cycle control is independent of differentiation, indicating that separate regulatory mechanisms must be involved in each process. The role of p57^{Kip2} appears to be restricted to progenitor cell cycle exit and it is rather the related p27^{Kip1} that prevents re-entry into the cycle of differentiated cells. While these data revealed a new transient intermediate between progenitors and differentiated cells, they also raised new questions and suggested that separate signals may control differentiation and cell cycle.

Copyright © 2010 S. Karger AG, Basel

Pituitary Stem Cells

The discovery of stem cells in many adult tissues led to their quest in the pituitary. Their existence was first supported by the isolation from the adult pituitary of a 'side population' (SP) of cells that are capable of forming clonal spheres in culture and that express markers of stem/progenitor cells [1]. Putative pituitary stem cells were further characterized both in the developing pituitary and adult tissue using a marker of many stem cells including embryonic stem cells, the *Sox2* gene [2]. It appears that *Sox2* expression marks a subpopulation of adult pituitary cells that can form spheres, self-renew and can be differentiated into different hormone-producing lineages. It was further suggested that when *Sox2*-expressing cells lose their self-renewal capability and presumably enter

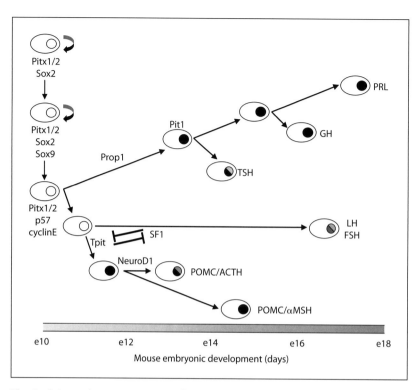

Fig. 1. Schematic representation of pituitary progenitors and differentiation sequence. All anterior and intermediate lobe pituitary cells that derive from Rathke's pouch are marked by expression of Pitx1/2. As discussed in the text, work from different groups has supported the idea that Sox2-positive pituitary stem cells can differentiate in all lineages and have the capacity for self renewal. These cells are thought to become Sox2 and Sox9 double-positive and may constitute a transit amplifying population of progenitors. These progenitors exit the cell cycle under the action of p57^{Kip2} to yield a transient population of non-cycling cells that also co-express cyclin E, but no marker of differentiation. The cell differentiation scheme is aligned with a temporal sequence for mouse pituitary development, progenitors being represented at the beginning of this sequence. Current evidence suggests that various progenitors and precursors are present throughout development but in decreasing numbers beyond the early fetal period; nonetheless, it is likely that similar cells/pathways are maintained throughout adulthood and involved in tissue maintenance.

the differentiation pathway, they switch on expression of *Sox9* and *nestin*. Both *Sox2* and *Sox9* are then extinguished upon terminal differentiation [2]. The use of a cell marking paradigm that relied on the activity of the *nestin* gene [3] and of a nestin-cre transgene supported a similar model of proliferating pituitary stem cells present in the adult gland that may serve as a pool for tissue maintenance [4]. It was indeed suggested that the adult pituitary renews a significant proportion of cells from its pool of stem cells. It was further suggested that upon entry into the differentiation pathway, these cells express regulators of normal pituitary development such as the Lhx3 and Lhx4 transcription factors and that their sequence of expression mimics their normal developmental sequence (fig. 1).

These studies have supported the model that renewal of pituitary cells in the adult gland depends on stem/progenitor cells that may represent the pool of proliferating cells in the adult, as they do in the developing pituitary [5]. Both during development and in adult, the bulk of undifferentiated proliferating cells are localized close to the cleft between anterior (AL) and neuro-intermediate (IL) lobes [2, 5]. These cells were suggested to reside in a 'niche' located along this cleft; cells of this putative 'niche' express the GDNF receptor α2 (GFRα2) and Prop1 [6]. These candidate stem/progenitor cells appear to express similar markers, such as Sox2, and thus similar cells and mechanisms may operate to expand pituitary cells in adult as in development.

Precursors and Differentiation

The different hormone-producing lineages of the pituitary differentiate sequentially during fetal development and organogenesis. We proposed a binary model of differentiation (fig. 1) starting from multipotent stem cells that first enter the differentiation pathway as multivalent progenitors leading to differentiated cells expressing a single hormone-coding gene [7, 8]. Whereas studies of cell proliferation in the pituitary indicated that hormone-expressing differentiated cells divide rarely, it appeared that it is primarily undifferentiated cells that proliferate. However, the lack of appropriate molecular markers has made it impossible to clearly define the nature of proliferating cells, both in development and adult: thus, the proliferating cells may include a mixture of true multipotent stem cells and/or various progenitors engaged in one or the other sublineages such as the putative common precursor for the POMC and gonadotrope lineages, and the progenitors of the somatolactotrope lineages (fig. 1). In addition to studies in normal tissues, the analysis of pituitary cell proliferation following end-organ ablation such as after adrenalectomy and/or gonadectomy, has shown that it is mostly hormone-negative cells that proliferate in these conditions [9]. It is noteworthy that a comparison of the pool of cells that expand following adrenalectomy together with gonadectomy appeared to be the same as following either adrenalectomy or gonadectomy [9], clearly supporting the notion that a common pool of progenitors/precursors is responsible for these responses (fig. 1). Nonetheless, it cannot be excluded at this juncture that slowly dividing differentiated cells may contribute to tissue maintenance.

The control of differentiation became better understood with the discovery over the last decade of critical transcription factors directing this process. Thus, the Pit1 transcription factor was the first such regulator identified for its role in differentiation of the GH, PRL and TSH lineages [10, 11]. A similar and critical role was shown for the Tbox factor Tpit that is required for differentiation of the POMC lineages, both AL corticotropes and IL melanotropes [8, 12, 13]. A gonadotrope-specific function was shown for the orphan nuclear receptor SF1 but this function appears to act late in differentiated gonadotropes and be dependent on hypothalamic GnRH signaling

[14, 15]. Also, GATA-2 which is expressed fairly broadly in early pituitary organogenesis contributes to the function of gonadotrope/thyrotrope cells [16].

These various studies have supported a model of pituitary cell differentiation in which progenitors and precursors become sequentially more restricted in their potential through expression of these transcription factors. The action of some of these factors was shown to be mutually exclusive and thus may serve to ensure differentiation into one lineage at the exclusion of another with which it shares a common precursor. For example, our analysis of Tpit-positive cells during pituitary development led us to identify a very small number of cells (5–20 cells per pituitary) that appeared positive for both Tpit and SF1, and we have shown transcriptional antagonism between these two factors such that the activity of one will ultimately predominate over the other [8]. These observations are consistent with the existence of rare cells positive for POMC and αGSU [17]. Since it appears that cell fate decisions may be established in progenitors after their exit from the cell cycle, the relationship between progenitor cycle exit and differentiation may thus be critical to set these processes in motion.

Cell Cycle Exit in Early Development

Early pituitary organogenesis includes a period of intense cell proliferation starting from the Rathke's pouch stage around embryonic day 10.5 (e10.5) of mouse development till about e14.5 when cell division slows down progressively towards the adult tissue where dividing cells are only very rarely observed. Throughout development and in the adult gland, the bulk of proliferation occurs around the cleft separating the IL from AL [2, 5] and it is along this pituitary cleft that Sox2-positive and nestin-positive progenitors have been observed in the adult gland [2, 4]. Thus, in early development, the proliferating cells are mostly observed along the pituitary cleft whereas the first differentiated cells (Tpit and ACTH-positive corticotropes) appear on the ventral side of the developing gland. Later, gonadotropes also first appear on the ventral side of the developing AL whereas the Pit1-dependent lineages first appear more dorsally within the center of the developing AL [18]. It is striking that at early stages of development (e12.5–e14.5), cells that are located between the proliferating dorsal cells and the ventral differentiated cells are not labelled by markers of either proliferation or differentiation. This observation led us to investigate expression of cell cycle markers in order to define regulatory factors that may be involved in cell cycle exit of pituitary progenitors. These studies led to the striking realization that pituitary progenitors first exit the cell cycle to transit through an intermediate state before switching-on differentiation markers such as Tpit or Pit1 [19]. It was also interesting to find that these transient undifferentiated cells are uniquely marked by expression of the cell cycle inhibitor (cdki), p57^{Kip2} and by cyclin E (fig. 2). Neither of these molecular markers is detectable by immunohistochemistry at any other time in the life of pituitary cells starting from Rathke's

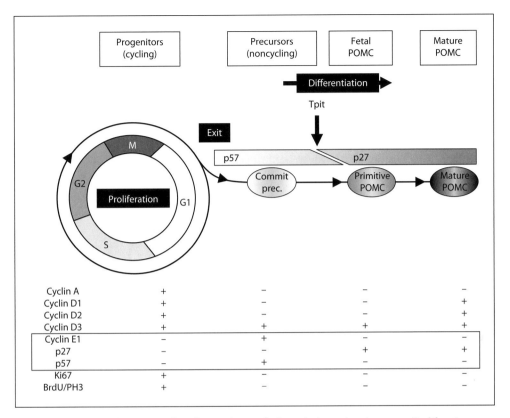

Fig. 2. Expression of various cell cycle regulators during pituitary development. Proliferating progenitors are most abundant in the early fetal pituitary (e12–e14) and they are labeled by the cell proliferation markers, BrdU, phospho-histone H3 (PH3) and Ki67. They also express different cyclins. Concomitantly, a group of noncycling cells (negative for proliferation markers and labeled 'Precursors' in the figure) are uniquely marked by expression of the cell cycle inhibitor p57^{Kip2} and by cyclin E; these cells are located between cycling progenitors present along the pituitary cleft and differentiated cells found on the ventral side of the developing gland. These markers only label this transient cell population and they are not detected in differentiated cells, such as in POMC- and Tpit-positive corticotrope cells. Differentiation into this and other (not shown) lineages is accompanied by expression of another Cip/Kip inhibitor p27^{Kip1}. p27^{Kip1} expression is maintained in all adult differentiated pituitary cells and its knockout leads to pituitary tumor development.

pouch through late adult life. The clear temporal separation between progenitor cell cycle exit and differentiation suggests that these transient undifferentiated cells may represent a critical stage in organogenesis during which cells are subjected to differentiation cues and that it may be during this phase that cell fates (and proportion) of each lineage are decided.

Since p57^{Kip2} is the main Cip/Kip inhibitor expressed in early development [19] (fig. 2), it was hypothesized that p57^{Kip2} may be the major cell cycle regulator driving proliferating progenitors out of the cell cycle. This hypothesis was ascertained in *p57^{Kip2}* knockout mice and indeed the pituitaries of these mice are hyperplastic and have

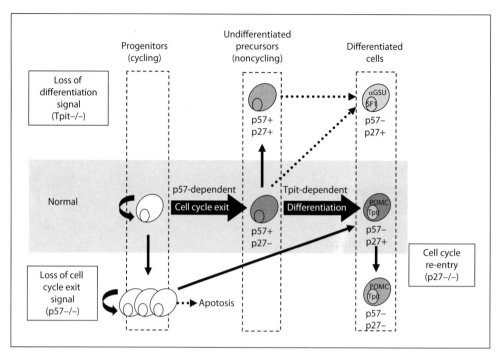

Fig. 3. Control of pituitary cell proliferation by the Cip/Kip inhibitors p57^{Kip2} and p27^{Kip1}. This scheme highlights the role of p57^{Kip2} that drives cell cycle exit of pituitary progenitors, yielding a transient population of undifferentiated precursors. Upon differentiation, these cells lose expression of p57^{Kip2} and become p27^{Kip1}-positive, as shown for the Tpit-dependent (corticotrope) POMC cells. In *p57^{Kip2}–/–* mutant pituitaries, proliferating progenitors accumulate until late fetal development when many cells undergo apoptosis; a fraction of cells nonetheless undergo differentiation and these differentiated cells express p27^{Kip1}. In agreement with the idea that p27^{Kip1} prevents re-entry into cycle of differentiated pituitary cells, cells from pituitaries of *p57^{Kip2}–/–;p27^{Kip1}–/–* mice never exit the cell cycle. In *Tpit–/–* mutant mice, differentiation into the POMC lineages (corticotrope and melanotrope) is largely prevented and this leads to accumulation of p57^{Kip2} and p27^{Kip1} double-positive cells that resemble undifferentiated precursors. A small fraction of those cells differentiate into the expected POMC cell fate whereas a greater number switches cell fate to become gonadotropes: in either case, cells that do differentiate lose expression of p57^{Kip2} and are only p27^{Kip1}-positive, like normal differentiated pituitary cells.

more proliferating progenitors; but, surprisingly, failure of early cell cycle exit is not definitive since a proportion of cells do eventually exit the cell cycle and differentiate. Thus, cell differentiation does not absolutely require p57^{Kip2}-dependent cell cycle exit. However although some cells are rescued, exit the cell cycle and differentiate, a large number of cells appear to be misprogrammed since they undergo apoptosis in late fetal development starting at e17.5 [19]. This apoptotic response clearly suggests that the p57^{Kip2}-dependent mechanism of cell cycle exit plays a unique role in progression of pituitary cells from progenitor to differentiation during development (fig. 3). This p57^{Kip2}-dependent transient state may be required to complete re-programming of cells from a 'stem' to 'differentiation' status, and may involve processes such as chromatin modification and re-modeling.

Blockade of Cell Cycle Re-Entry in Differentiated Cells

The absence of p57^{Kip2} does not prevent all cell cycle exit in the developing pituitary of *p57^{Kip2}–/–* mice because p27^{Kip1}, another member of the Cip/Kip family of cdki, is expressed in differentiated pituitary cells. The onset of p27^{Kip1} expression coincides with the appearance of differentiation markers such as Tpit [19] and it appears to slightly precede differentiation since a small number of p57^{Kip2}-positive/ p27^{Kip1}-positive/ Tpit-negative cells are observed whereas all Tpit-positive cells are p27^{Kip1}-positive (fig. 2). The onset of p27^{Kip1} is not altered in absence of p57^{Kip2}, clearly indicating that p57^{Kip2} is not the only regulator of p27^{Kip1} expression. However, p27^{Kip1} switches on in a fraction of cells in *p57^{Kip2}–/–* pituitaries and this may provide an explanation why this subset of cells appears to differentiate normally. But it is still an open question as to why other cells fail to switch-on p27^{Kip1} and undergo apoptosis.

These observations clearly suggested that cell cycle progression of differentiated cells is under the control of p27^{Kip1}. This hypothesis was assessed in *p27^{Kip1}–/–* mice that ultimately will develop pituitary tumors [20–23]. During the fetal period, *p27^{Kip1}–/–* pituitaries exhibit a few proliferating differentiated cells, something that is almost never observed in normal pituitary tissue [19]. Otherwise, the pituitaries of these animals have quite normal appearance and cell composition; it takes a few months for progression to clear tumor formation. The hypothesis that p57^{Kip2} and p27^{Kip1} may collectively control cell cycle exit of progenitors and differentiated cells, respectively, was assessed in double knockout *p57^{Kip2}–/–; p27^{Kip1}–/–* pituitaries. Strikingly and in clear support of our model (fig. 3), cells of these pituitaries are all in proliferation and the gland is hyperplastic similar to the *p57^{Kip2}–/–* pituitaries [19]. These studies clearly support a model in which two Cip/Kip inhibitors are primarily responsible for control of cell cycle progression in normal development: p57^{Kip2} drives proliferating progenitors out of the cycle and cell cycle re-entry is prevented in differentiated cells by the action of p27^{Kip1} (fig. 3).

The knockouts of other cell cycle inhibitors have suggested that they play an important role in restricting cell proliferation of pituitary cells since these knockouts led to the occurrence of pituitary tumors. These include knockouts of the INK4 family of cdki's, such as p18^{INK4c} [23, 24]. Our preliminary analyses suggest that these other cdki's are not expressed at very high levels during fetal pituitary development and/or have post-natal onsets of expression. This includes p18^{INK4c} and the third member of the Cip/Kip family, p21^{Cip1}. Hence, the model of cell cycle control being exerted by p57^{Kip2} and p27^{Kip1} during fetal organogenesis is not incompatible with the roles of other cdki's later in adult pituitary.

Independent Control of Differentiation and Cell Cycle

It is often believed and said that differentiation and proliferation appear mutually exclusive in development such that when differentiation is triggered, it is accompanied by cell cycle exit. The pituitary appeared to follow this belief since the wave of cell proliferation precedes differentiation and the fully differentiated adult pituitary exhibits very little cell division. However, our recent work has clearly indicated that cell cycle exit of progenitors is the first in a series of events leading to the establishment of the differentiated cell lineages of the adult pituitary. As discussed above, progenitor cell cycle exit first leads to a transient state in which precursors are no longer in proliferation but do not yet express markers of terminal cell differentiation (fig. 3). The only clue for the importance of this transition state is provided by the apoptotic response observed in absence of $p57^{Kip2}$. Further, analyses of $p57^{Kip2}-/-$ and $p27^{Kip1}-/-$ mice that are defective in pituitary cell cycle exit, clearly showed that cell differentiation is independent of cell cycle exit or re-entry. This was shown for cells of lineages marked by Tpit (corticotrope), Pit1 (somatotrope and melanotrope) and SF1 (gonadotrope) [19]. The importance of differentiation itself for control of cell cycle was investigated in $Tpit-/-$ pituitaries that are blocked in corticotrope and melanotrope differentiation [8, 13]. Since the $Tpit$ knockout retains only a very small number of POMC-positive cells in the AL, these analyses were more easily carried out in the IL melanotropes. The IL cells of $Tpit-/-$ mice have a further interest in relation to the question of differentiation versus proliferation since a fraction of those cells change cell fate to become gonadotropes expressing αGSU, βLH and βFSH [8]. Interestingly, the absence of $p57^{Kip2}$ did not prevent differentiation into melanotropes or the more frequent cell fate change into gonadotropes. Consistent with observations in normal pituitaries, the $Tpit-/-$ cells that differentiate into either fate, switch-off $p57^{Kip2}$ and express $p27^{Kip1}$. However, the bulk of the $Tpit-/-$ IL cells that do not differentiate, retain expression of $p57^{Kip2}$ in addition to the normal onset of $p27^{Kip1}$ expression, i.e. a significant proportion of those cells co-express $p57^{Kip2}$ and $p27^{Kip1}$, contrary to normal. These observations suggest that differentiation (either in melanotrope or gonadotrope fate) ensures extinction of $p57^{Kip2}$ expression but that expression of $p27^{Kip1}$ is independent of these processes. This observation further stresses the mystery around the biological importance of the $p57^{Kip2}$-positive transient state since from the strict perspective of cell cycle exit, the onset of $p27^{Kip1}$ should suffice to do the job.

Perspective

In summary, it appears that cell cycle control exerted by the Cip/Kip cdki's $p27^{Kip1}$ and $p57^{Kip2}$ does not directly impact pituitary cell differentiation in any way whereas the differentiation process into either melanotrope and gonadotrope fates, leads to switch-off of $p57^{Kip2}$ independently of any effect on $p27^{Kip1}$ expression. Taken collectively, these recent data indicate there is little relation between control of cell cycle progression and differentiation. Numerous signaling pathways have been involved over the last decade in the control of pituitary organogenesis and/or cell differentiation [25–30]: in view of the observed independence between cell cycle exit and cell fate decisions, it should now be interesting to re-assess many of the paradigms used to define the role of signaling pathways in pituitary development in order to find out whether some signals may not be preferentially acting on cell cycle progression rather than differentiation.

Acknowledgements

We are indebted to many colleagues from the laboratory for stimulating discussions. Work done in the author's laboratory was supported by grants from the Canadian Institutes of Health Research (CIHR) and the National Cancer Institute of Canada (NCIC).

References

1 Chen J, Hersmus N, Van DV, Caesens P, Denef C, Vankelecom H: The adult pituitary contains a cell population displaying stem/progenitor cell and early embryonic characteristics. Endocrinology 2005;146: 3985.

2 Fauquier T, Rizzoti K, Dattani M, Lovell-Badge R, Robinson IC: SOX2-expressing progenitor cells generate all of the major cell types in the adult mouse pituitary gland. Proc Natl Acad Sci USA 2008;105:2907.

3 Krylyshkina O, Chen J, Mebis L, Denef C, Vankelecom H: Nestin-immunoreactive cells in rat pituitary are neither hormonal nor typical folliculo-stellate cells. Endocrinology 2005;146:2376.

4 Gleiberman AS, Michurina T, Encinas JM, Roig JL, Krasnov P, Balordi F, Fishell G, Rosenfeld MG, Enikolopov G: Genetic approaches identify adult pituitary stem cells. Proc Natl Acad Sci USA 2008;105: 6332.

5 Ward RD, Raetzman LT, Suh H, Stone BM, Nasonkin IO, Camper SA: Role of PROP1 in pituitary gland growth. Mol Endocrinol 2005;19:698.

6 Garcia-Lavandeira M, Quereda V, Flores I, Saez C, Diaz-Rodriguez E, Japon MA, Ryan AK, Blasco MA, Dieguez C, Malumbres M, Alvarez CV: A GRFa2/Prop1/stem (GPS) cell niche in the pituitary. PLoS ONE 2009;4:e4815.

7 Drouin J: Molecular mechanisms of pituitary differentiation and regulation: implications for hormone deficiencies and for hormone resistance syndromes. Front Horm Res. Basel, Karger, 2006, vol 35, pp 74–87.

8 Pulichino AM, Vallette-Kasic S, Tsai JPY, Couture C, Gauthier Y, Drouin J: Tpit determines alternate fates during pituitary cell differentiation. Genes Dev 2003;17:738.

9 Nolan LA, Levy A: A population of non-luteinising hormone/non-adrenocorticotrophic hormone-positive cells in the male rat anterior pituitary responds mitotically to both gonadectomy and adrenalectomy. J Neuroendocrinol 2006;18:655.

10 Bodner M, Castrillo JL, Theill LE, Deerinck T, Ellisman M, Karin M: The pituitary-specific transcription factor GHF-1 is a homeobox-containing protein. Cell 1988;55:505.

11 Ingraham HA, Chen R, Mangalam HJ, Elsholtz HP, Flynn SE, Lin CR, Simmons DM, Swanson L, Rosenfeld MG: A tissue-specific transcription factor containing a homeodomain specifies a pituitary phenotype. Cell 1988;55:519.

12 Lamolet B, Pulichino AM, Lamonerie T, Gauthier Y, Brue T, Enjalbert A, Drouin J: A pituitary cell-restricted T-box factor, Tpit, activates POMC transcription in cooperation with Pitx homeoproteins. Cell 2001;104:849.

13 Pulichino AM, Vallette-Kasic S, Couture C, Gauthier Y, Brue T, David M, Malpuech G, Deal C, Van Vliet G, De Vroede M, Riepe FG, Partsch CJ, Sippell WG, Berberoglu M, Atasay B, Drouin J: Human and mouse Tpit gene mutations cause early onset pituitary ACTH deficiency. Genes Dev 2003;17:711.

14 Ingraham HA, Lala DS, Ikeda Y, Luo X, Shen WH, Nachtigal MW, Abbud R, Nilson JH, Parker KL: The nuclear receptor steroidogenic factor 1 acts at multiple levels of the reproductive axis. Genes Dev 1994;8:2302.

15 Zhao L, Bakke M, Krimkevich Y, Cushman LJ, Parlow AF, Camper SA, Parker KL: Steroidogenic factor 1 (SF1) is essential for pituitary gonadotrope function. Development 2001;128:147.

16 Charles MA, Saunders TL, Wood WM, Owens K, Parlow AF, Camper SA, Ridgway EC, Gordon DF: Pituitary-specific Gata2 knockout: effects on gonadotrope and thyrotrope function. Mol Endocrinol 2006;20:1366.

17 Pals K, Boussemaere M, Swinnen E, Vankelecom H, Denef C: A Pituitary cell type coexpressing messenger ribonucleic acid of proopiomelanocortin and the glycoprotein hormone alpha-subunit in neonatal rat and chicken: rapid decline with age and reappearance in vitro under regulatory pressure of corticotropin-releasing hormone in the rat. Endocrinology 2006;147:4738.

18 Japon MA, Rubinstein M, Low MJ: In situ hybridization analysis of anterior pituitary hormone gene expression during fetal mouse development. J Histochem Cytochem 1994;42:1117.

19 Bilodeau S, Roussel-Gervais A, Drouin J: Distinct developmental roles of cell cycle inhibitors p57^{Kip2} and p27^{Kip1} distinguish pituitary progenitor cell cycle exit from cell cycle re-entry of differentiated cells. Mol Cell Biol 2009;29:1895.

20 Fero ML, Rivkin M, Tasch M, Porter P, Carow CE, Firpo E, Polyak K, Tsai LH, Broudy V, Perlmutter RM, Kaushansky K, Roberts JM: A syndrome of multiorgan hyperplasia with features of gigantism, tumorigenesis, and female sterility in p27(Kip1)-deficient mice. Cell 1996;85:733.

21 Kiyokawa H, Kineman RD, Manova-Todorova KO, Soares VC, Hoffman ES, Ono M, Khanam D, Hayday AC, Frohman LA, Koff A: Enhanced growth of mice lacking the cyclin-dependent kinase inhibitor function of p27(Kip1). Cell 1996;85:721.

22 Nakayama K, Ishida N, Shirane M, Inomata A, Inoue T, Shishido N, Horii I, Loh DY, Nakayama K: Mice lacking p27(Kip1) display increased body size, multiple organ hyperplasia, retinal dysplasia, and pituitary tumors. Cell 1996;85:707.

23 Franklin DS, Godfrey VL, Lee H, Kovalev GI, Schoonhoven R, Chen-Kiang S, Su L, Xiong Y: CDK inhibitors p18(INK4c) and p27(Kip1) mediate two separate pathways to collaboratively suppress pituitary tumorigenesis. Genes Dev 1998;12:2899.

24 Latres E, Malumbres M, Sotillo R, Martin J, Ortega S, Martin-Caballero J, Flores JM, Cordon-Cardo C, Barbacid M: Limited overlapping roles of P15(INK4b) and P18(INK4c) cell cycle inhibitors in proliferation and tumorigenesis. EMBO J 2000;19:3496.

25 Treier M, Gleiberman AS, O'Connell SM, Szeto DP, McMahon JA, McMahon AP, Rosenfeld MG: Multistep signaling requirements for pituitary organogenesis in vivo. Genes Dev 1998;12:1691.

26 Treier M, O'Connell S, Gleiberman A, Price J, Szeto DP, Burgess R, Chuang PT, McMahon AP, Rosenfeld MG: Hedgehog signaling is required for pituitary gland development. Development 2001;128:377.

27 Ericson J, Norlin S, Jessell TM, Edlund T: Integrated FGF and BMP signaling controls the progression of progenitor cell differentiation and the emergence of pattern in the embryonic anterior pituitary. Development 1998;125:1005.

28 Raetzman LT, Ross SA, Cook S, Dunwoodie SL, Camper SA, Thomas PQ: Developmental regulation of Notch signaling genes in the embryonic pituitary: Prop1 deficiency affects Notch2 expression. Dev Biol 2004;265:329.

29 Brinkmeier ML, Potok MA, Cha KB, Gridley T, Stifani S, Meeldijk J, Clevers H, Camper SA: TCF and Groucho-related genes influence pituitary growth and development. Mol Endocrinol 2003;17: 2152.

30 Zhu X, Zhang J, Tollkuhn J, Ohsawa R, Bresnick EH, Guillemot F, Kageyama R, Rosenfeld MG: Sustained Notch signaling in progenitors is required for sequential emergence of distinct cell lineages during organogenesis. Genes Dev 2006;20:2739.

Jacques Drouin
Laboratoire de génétique moléculaire, Institut de recherches cliniques de Montréal (IRCM)
110, avenue des Pins Ouest
Montréal QC, H2W 1R7 (Canada)
Tel. +1 514 987 5680, Fax +1 514 987 5575, E-Mail jacques.drouin@ircm.qc.ca

Role of Estrogens in Anterior Pituitary Gland Remodeling during the Estrous Cycle

S. Zárate · V. Zaldivar · G. Jaita · L. Magri · D. Radl · D. Pisera · A. Seilicovich

Instituto de Investigaciones en Reproducción, Facultad de Medicina, Universidad de Buenos Aires, Buenos Aires, Argentina

Abstract

In this review, we analyze the action of estrogens leading to the remodeling of the anterior pituitary gland, especially during the estrous cycle. Proliferation and death of anterior pituitary cells and especially lactotropes is regulated by estrogens, which act by sensitizing these cells to both mitotic and apoptotic stimuli such as TNF-α, FasL and dopamine. During the estrous cycle, the changing pattern of gonadal steroids is thought to modulate both cell proliferation and death in the anterior pituitary gland, estrogens being key players in cell turnover. The mechanisms involved in estrogen-modulated cell renewal in the anterior pituitary gland during the estrous cycle could include an increase in the expression of proapoptotic cytokines as well as the increase in the Bax/Bcl-2 ratio at proestrus, when estrogen levels are highest and a peak of apoptosis, in particular of lactotropes, is evident in this gland. Estrogens exert rapid antimitogenic and proapoptotic actions in the anterior pituitary through membrane-associated estrogen receptors, a mechanism that might also be involved in remodeling of this gland during the estrous cycle.

Copyright © 2010 S. Karger AG, Basel

Anterior Pituitary Cell Renewal

The anterior pituitary gland undergoes processes of remodeling in several physiological conditions such as pregnancy, lactation and the estrous cycle [reviewed in 1]. For example, remodeling of the anterior pituitary gland after termination of lactation occurs through apoptosis and involves changes in the expression of Bax and Bcl-2, a proapoptotic and an antiapoptotic protein, respectively, of the Bcl-2 family [2].

In basal conditions, the normal, adult male pituitary gland has a parenchymal cell turnover >1.5% per day [3], while anterior pituitary mitotic activity in cycling females doubles that in males overall [4]. Estrogens are recognized modulators of pituitary cell renewal, sensitizing cells to mitogenic and apoptotic signals [1, 4–7]. Lactotropes are the anterior pituitary cell subpopulation with the highest turnover during the estrous cycle, proliferating at estrus and showing high sensitivity to proapoptotic stimuli at proestrus, when estrogen levels are highest [1]. In addition to the well known proliferative effects of estrogens on lactotropes [1], recent evidence has shown that estrogens also trigger antiproliferative responses in anterior pituitary cells, inhibiting lactotrope proliferation induced by growth factors and insulin [1, 8, 9]. Also, estrogens were recently shown to sensitize lactotropes to the proapoptotic action of both dopamine and the specific D2 receptor agonist cabergoline, indicating that estrogens are required to induce apoptosis initiated by D2 receptor activation in these cells [10].

Estrogens trigger the action of death receptor ligands such as TNF-α and FasL, inducing apoptosis of lactotropes and somatotropes. In fact, TNF-α and Fas receptor activation increases the percentage of apoptotic lactotropes and somatotropes only when anterior pituitary cells from ovariectomized rats are cultured with 17β-estradiol E2 [1, 5, 6, 11]. Therefore, the apoptotic effect induced by these cytokines in anterior pituitary cells from female rats is estrogen-dependent and observed in rats at proestrus but not at diestrus [5, 11].

Estrogens per se also induce apoptosis of anterior pituitary cells. We reported that E2 treatment increases the apoptotic rate in the anterior pituitary gland of ovariectomized rats [7, 12]. In vitro studies showed that this effect may result, at least in part, from direct action of estrogens on anterior pituitary cells [6, 7]. Also, E2 per se increased the percentage of apoptotic somatotropes from ovariectomized rats in culture [6].

Although progesterone does not intrinsically affect apoptosis of anterior pituitary cells, it antagonizes the permissive action of E2 on TNF-α- and FasL-induced apoptosis of lactotropes and somatotropes [1, 6, unpubl. results].

Mechanisms of Estradiol Action in Anterior Pituitary Cell Turnover

The mechanisms by which estrogens sensitize anterior pituitary cells, in particular lactotropes and somatotropes, to proapoptotic stimuli are beginning to be elucidated. One mechanism involved in this estrogenic action is increased expression of cytokines. In fact, we observed that expression of both FasL and Fas receptor in anterior pituitary cells is estrogen-dependent and higher in anterior pituitary cells from rats killed at proestrus than at diestrus [11]. Estrogens also stimulate TNF-α expression, an effect perhaps involved in the higher release of TNF-α in the anterior pituitary gland from rats at proestrus [13]. Thus, the steroid environment in the morning of proestrus could increase local production of auto/paracrine mediators which would lead to apoptosis of anterior pituitary cells at this stage of the estrous cycle.

Apoptosis is executed by a family of proteases called caspases, which may be activated either by cell-surface death receptors or by disturbance of the mitochondrial membrane. The integrity of mitochondrial membranes is controlled by a balance between the antagonistic actions of the proapoptotic and antiapoptotic members of the Bcl-2 protein family [14]. The apparently paradoxical effect of estrogens on apoptosis in different cell types [15, 16] could result from their ability to differentially modulate the expression of proapoptotic and antiapoptotic proteins of the Bcl-2 family [17]. The Bax/Bcl-2 ratio increases in the anterior pituitary gland at proestrus, coincidently with enhanced apoptosis, suggesting that changes in the expression of these proteins may be involved in regulation of apoptosis in this gland during the estrous cycle [7]. Increase in the Bax/Bcl-2 ratio and decreased expression of the antiapoptotic protein Bcl-xL at proestrus may result, at least in part, from direct action of estrogens on anterior pituitary cells [7]. Also, estradiol was found to enhance expression of the proapoptotic protein Bad in lactotropes [18]. Although expression of a dominant negative form of ERα in the anterior pituitary cell line GH4 induces apoptosis, the overexpression of wild type ERα also induces apoptosis and decreases Bcl-2 expression in this cell line [19]. Therefore, the increase in proapoptotic vs. antiapoptotic proteins of the Bcl-2 family could be involved in estrogen sensitizing to proapoptotic stimuli in anterior pituitary cells.

Rapid Actions of Estrogens in Anterior Pituitary Cells

Estrogens exert their multiple actions by binding to estrogen receptors ERα and ERβ, members of the nuclear receptor superfamily of ligand-activated transcription factors [20]. Upon ligand binding, ERs undergo conformational changes which lead to their dimerization, recruitment of coactivator and corepressor molecules and binding to target genes to either activate or repress gene transcription. Also, ligand-bound ERs can tether other transcription factors, such as SP1, c-Fos/c-JunB (AP-1) and NFκB, and thus regulate gene transcription independently of direct binding to DNA. These events are described as nuclear-initiated steroid signaling [21]. However, E2 action may also be exerted through membrane-initiated E2 pathways. In these extra-nuclear steroid pathways, E2 binds to membrane-associated ERs, thereby rapidly activating different second messengers which, in turn, regulate the activity of different proteins. In this way, by changing the phosphorylation status of some proteins, E2 can influence gene transcription indirectly [20, 21]. An additional ER pathway involves crosstalk between ERs and growth factor receptors, during which gene activation occurs in the absence of ligand through second messenger pathways resulting in altered phosphorylation of ERs [20, 21].

E2 rapidly stimulates signaling pathways in anterior pituitary cells, leading to ion flux and hormone release [22, 23]. E2 has been reported to induce rapid release of prolactin from anterior pituitary cells from both male and female rats, an action mimicked by the membrane-impermeant E2-albumin conjugate, E2-BSA [9, 22]. In the prolactinoma cell line GH3/B6/F10 with high expression of membrane receptor ERα, E2 and E2-peroxidase (another cell-impermeable analogue of E2) induce a rapid increase in intracellular calcium levels and prolactin release [23]. Extracellular-regulated kinase (ERK) activation was suggested to be involved in these E2 actions [24]. E2 also exerts rapid antimitogenic and even proapoptotic actions in anterior pituitary cells, probably through membrane-associated ERα [9, 25]. Gutierrez et al. reported that early mitogenic activity promoted by insulin on lactotropes was abrogated by E2 or E2-BSA, an action involving PKC, ERK1/2, Pit-1 and probably mERα [9]. Recently, we showed that both E2 and E2-BSA rapidly induce apoptosis of total anterior pituitary cells, lactotropes and somatotropes, at least in part, by a caspase-dependent mechanism [25]. The proapoptotic effect of E2-BSA on anterior pituitary cells is abrogated by the pure antagonist of classical ERs, ICI 182,780, indicating that this estrogenic action is mediated by activation of membrane-associated receptors and that these receptors are likely classical ERs associated to the plasma membrane [25].

In addition to ERα and ERβ, several variant isoforms of ER mRNA are produced in adult rat pituitary by alternative mRNA splicing or base insertions within an exon [26]. These variants may act alone or alter the function of full-length ERs. Also, adult female rat pituitary expresses a unique truncated form of ERα, TERP-1, containing the C-terminal region of full-length ERα. TERP-1 expression varies along the estrous cycle and peaks at proestrous, exceeding ERα expression levels, suggesting a possible regulatory role for this receptor variant in physiological conditions. TERP-1 cannot bind DNA but modulates effects of full-length ERα and ERβ through protein-protein interactions [26]. The nature of receptor/s mediating non canonical E2 actions is still elusive. Current evidence favors the idea that nuclear (classical) and membrane-associated ER are the same protein [27]. However, truncated variants of ERα, G protein coupled receptors, such as GPR30, and even other proteins have also been implicated in membrane-initiated estrogen signaling [21].

The anterior pituitary of adult female rats expresses mainly ERα, although ERβ is expressed in nearly every adult pituitary cell type, albeit at a lower level [28]. ERα but not ERβ protein levels vary along the estrous cycle with an expression pattern positively related to circulating estradiol levels [28]. In the rat, lactotropes are the population with the highest immunoreactivity for ERα, followed by somatotropes [28]. TERP-1 was found in ER-containing populations, particularly lactotropes and gonadotropes [26]. At the plasma membrane level, estrogen binding sites were observed in dispersed anterior pituitary cells, with a patchy distribution on cell surface [22]. mERα was detected both in rat pituitary tumor cell line GH3/B6/F10 [29] and in normal anterior pituitary lactotropes, while mERβ was not found in these cells [9]. By using a flow cytometry approach which preserves the integrity of anterior pituitary cell membrane, mERα was detected in the plasma membrane of lactotropes and somatotropes [25]. Since the antibody used was directed to an epitope within the C-terminal region of the classical ERα, the mERα detected could involve full-length ERα and/or TERP-1. About 40 % of lactotropes and approximately 13 % of somatotropes express mERα [25]. Since levels of mER were suggested to be controlled more dynamically than its nuclear counterpart [30], the presence of mERs in anterior pituitary cells and especially in lactotropes would allow them to change their sensitivity to different stimuli depending on the pattern of circulating gonadal hormones, as occurs during the estrous cycle. In fact, we detected a higher number of lactotropes expressing mERα at proestrous than at diestrus, whereas the number of somatotropes expressing this membrane receptor was not altered at these two stages of the estrous cycle [unpubl. results]. E2 increased the number of mERα-expressing lactotropes while progesterone abrogated this estrogenic effect [unpubl. results]. It was suggested that intracellular and membrane fractions of ERα might be part of the same ER pool showing a different balance of subcellular distribution [31], and gonadal steroids may be key regulators of this differential location. In fact, it has been shown that E2 causes trafficking of ERα to the plasma membrane within minutes in PC-12 cells [32].

Concluding Remarks

Evidence shows that cell proliferation and apoptosis are coupled processes. Activation of cell proliferation necessarily primes the cellular apoptotic program that, if not abrogated by appropriate survival signals, removes the affected cell [33]. Gonadectomy followed by testosterone treatment was shown to induce concurrent waves of mitosis and apoptosis in the anterior pituitary which are closely linked [34]. Estrogens seem to tightly modulate pituitary cell renewal by sensitizing cells to both mitogenic stimuli and proapoptotic signals, thereby maintaining tissue homeostasis. The presence of mERs in lactotropes and somatotropes would allow them to rapidly and accurately assess changes in hormone levels. During the estrous cycle, activation of mERs could be a mechanism for responding rapidly to the changing environment. About 2.5% of lactotropes proliferate at estrus and therefore a similar number of these cells might die to maintain the size of the lactotrope subpopulation. Although it remains uncertain why females have more mitosis at estrus and more apoptosis at proestrus, it is possible to speculate that lactotrope proliferation at estrus prepares this pituitary subpopulation to meet needs during pregnancy. When pregnancy does not occur, lactotropes might die to compensate the increased number of these cells. Coupled to concurrent reduction in the prevalence of mitosis, increased cell death at proestrus seems likely to contribute to maintenance of homeostasis of the pituitary gland. Despite the low percentage of anterior pituitary cells renewed in each estrous cycle, dysregulation of this process could have considerable pathophysiological impact.

Acknowledgments

This work was funded by grants from the Agencia Nacional de Investigaciones Científicas y Tecnológicas, the Consejo Nacional de Investigaciones Científicas y Técnicas (CONICET) and the Universidad de Buenos Aires, Argentina.

References

1 Candolfi M, Zaldivar V, Jaita G, Seilicovich A: Anterior pituitary cell renewal during the estrous cycle. Front Horm Res. Basel, Karger, 2006, vol 35, pp 9–21.
2 Ahlbom E, Grandison L, Zhivotovsky B, Ceccatelli S: Termination of lactation induces apoptosis and alters the expression of the Bcl-2 family members in the rat anterior pituitary. Endocrinology 1998;139: 2465–2471.
3 Nolan LA, Kavanagh E, Lightman SL, Levy A: Anterior pituitary cell proliferation control: basal cell turnover and the effects of adrenalectomy and dexamethasone treatment. J Neuroendocrinol 1998; 10:207–215.
4 Oishi Y, Okuda M, Takahashi H, Fujii T, Morii S: Cellular proliferation in the anterior pituitary gland of normal adult rats: influences of sex, estrous cycle, and circadian change. Anat. Rec 1993;235:111–120.
5 Candolfi M, Zaldivar V, De Laurentiis A, Jaita G, Pisera D, Seilicovich A: TNF-alpha induces apoptosis of lactotropes from female rats. Endocrinology 2002;143:3611–3617.
6 Candolfi M, Jaita G, Zaldivar V, Zarate S, Ferrari L, Pisera D, Castro MG, Seilicovich A: Progesterone antagonizes the permissive action of estradiol on tumor necrosis factor-alpha-induced apoptosis of anterior pituitary cells. Endocrinology 2005;146: 736–743.
7 Zaldivar V, Magri ML, Zárate S, Jaita G, Eijo G, Radl D, Ferraris J, Pisera D, Seilicovich A: Estradiol increases the Bax/Bcl-2 ratio and induces apoptosis in anterior pituitary cells. Neuroendocrinology 2009; 90:292–300.
8 Kawashima K, Yamakawa K, Takahashi W, Takizawa S, Yin P, Sugiyama N, Kanba S, Arita J: The estrogen-occupied estrogen receptor functions as a negative regulator to inhibit cell proliferation induced by insulin/IGF-1: a cell context-specific antimitogenic action of estradiol on rat lactotrophs in culture. Endocrinology 2002;143:2750–2758.
9 Gutiérrez S, De Paul AL, Petiti JP, del Valle Sosa L, Palmeri CM, Soaje M, Orgnero EM, Torres AI: Estradiol interacts with insulin through membrane receptors to induce an antimitogenic effect on lactotroph cells. Steroids 2008;73:515–527.
10 Radl DB, Zárate S, Jaita G, Ferraris J, Zaldivar V, Eijo G, Seilicovich A, Pisera D: Apoptosis of lactotrophs induced by D2 receptor activation is estrogen dependent. Neuroendocrinology 2008;88:43–52.
11 Jaita G, Candolfi M, Zaldivar V, Zárate S, Ferrari L, Pisera D, Castro MG, Seilicovich A: Estrogens up-regulate the Fas/FasL apoptotic pathway in lactotropes. Endocrinology 2005;146:4737–4744.
12 Pisera D, Candolfi M, Navarra S, Ferraris J, Zaldivar V, Jaita G, Castro MG, Seilicovich A: Estrogens sensitize anterior pituitary gland to apoptosis. Am J Physiol Endocrinol Metab 2004;287:767–771.
13 Theas S, Pisera D, Duvilanski B, De Laurentiis A, Pampillo M, Lasaga M, Seilicovich A: Estrogens modulate the inhibitory effect of tumor necrosis factor-alpha on anterior pituitary cell proliferation and prolactin release. Endocrine 2000;12:249–255.
14 Kroemer G, Galluzzi L, Brenner C: Mitochondrial membrane permeabilization in cell death. Physiol Rev 2007;87:99–163.
15 Manolagas SC, Kousteni S, Jilka RL: Sex steroids and bone. Recent Prog Horm Res 2002;57:385–409.
16 Lewis JS, Meeke K, Osipo C, Ross EA, Kidawi N, Li T, Bell E, Chandel NS, Jordan VC: Intrinsic mechanism of estradiol-induced apoptosis in breast cancer cells resistant to estrogen deprivation. J Natl Cancer Inst 2005;97:1746–1759.
17 Chao DT, Korsmeyer SJ: BCL-2 family: regulators of cell death. Annu Rev Immunol 1998;16:395–419.
18 Kulig E, Camper SA, Kuecker S, Jin L, Lloyd RV: Remodeling of hyperplastic pituitaries in hypothyroid us-subunit knockout mice after thyroxine and 17β-estradiol treatment: role of apoptosis. Endocr Pathol 1998;9:261–274.
19 Lee EJ, Duan WR, Jakacka M, Gehm BD, Jameson JL: Dominant negative ER induces apoptosis in GH(4) pituitary lactotrope cells and inhibits tumor growth in nude mice. Endocrinology 2001;142:3756–3763.
20 McDevitt MA, Glidewell-Kenney C, Jimenez MA, Ahearn PC, Weiss J, Jameson JL, Levine JE: New insights into the classical and non-classical actions of estrogen: evidence from estrogen receptor knock-out and knock-in mice. Mol Cell Endocrinol 2008; 290:24–30.
21 Song RX, Santen RJ: Membrane initiated estrogen signaling in breast cancer. Biol Reprod 2006;75:9–16.
22 Christian HC, Morris JF: Rapid actions of 17beta-oestradiol on a subset of lactotrophs in the rat pituitary. J Physiol 2002;539:557–66.
23 Bulayeva NN, Wozniak AL, Lash LL, Watson CS: Mechanisms of membrane estrogen receptor-alpha-mediated rapid stimulation of Ca^{2+} levels and prolactin release in a pituitary cell line. Am J Physiol Endocrinol Metab 2005;288:388–397.
24 Bulayeva NN, Gametchu B, Watson CS: Quantitative measurement of estrogen-induced ERK 1 and 2 activation via multiple membrane-initiated signaling pathways. Steroids 2004;69:181–92.

25 Zárate S, Jaita G, Zaldivar V, Radl D, Eijo G, Ferraris J, Pisera D, Seilicovich A: Estrogens exert a rapid apoptotic action in anterior pituitary cells. Am J Physiol Endocrinol Metab 2009;296:E664–E671.

26 Shupnik MA: Oestrogen receptors, receptor variants and oestrogen actions in the hypothalamic-pituitary axis. J Neuroendocrinol 2002;14:85–94.

27 Razandi M, Pedram A, Park ST, Levin ER: Proximal events in signaling by plasma membrane estrogen receptors. J Biol Chem 2003;278:2701–2712.

28 González M, Reyes R, Damas C, Alonso R, Bello AR: Oestrogen receptor alpha and beta in female rat pituitary cells: an immunochemical study. Gen Comp Endocrinol 2008;155:857–868.

29 Watson CS, Campbell CH, Gametchu B: Membrane oestrogen receptors on rat pituitary tumour cells: immuno-identification and responses to oestradiol and xenoestrogens. Exp Physiol 1999;84:1013–1022.

30 Watson CS, Campbell CH, Gametchu B: The dynamic and elusive membrane estrogen receptor-alpha. Steroids 2002;67:429–437.

31 Zivadinovic D, Watson CS: Membrane estrogen receptor-alpha levels predict estrogen-induced ERK1/2 activation in MCF-7 cells. Breast Cancer Res 2005; 7:130–144.

32 Alyea RA, Laurence SE, Kim SH, Katzenellenbogen BS, Katzenellenbogen JA, Watson CS: The roles of membrane estrogen receptor subtypes in modulating dopamine transporters in PC-12 cells. J Neurochem 2008;106:1525–1533.

33 Evan G, Littlewood T: A matter of life and cell death. Science 1998;281:1317–1322.

34 Nolan LA, Levy A: The effects of testosterone and oestrogen on gonadectomised and intact male rat anterior pituitary mitotic and apoptotic activity. J Endocrinol 2006;188:387–396.

Adriana Seilicovich
Instituto de Investigaciones en Reproducción, Facultad de Medicina
Universidad de Buenos Aires, Paraguay 2155, piso 10
Buenos Aires, C1121ABG (Argentina)
Tel./Fax +54 11 48074052, E-Mail adyseili@fmed.uba.ar

Hyperprolactinemia following Chronic Alcohol Administration

Dipak K. Sarkar

Endocrine Program, Biomedical Division of the Center of Alcohol Studies and Department of Animal Sciences, Rutgers, The State University of New Jersey, New Brunswick, N.J., USA

Abstract

There are several reports showing evidence for the existence of high levels of prolactin (PRL) in alcoholic men and women. Alcohol-induced hyperprolactinemia has also been demonstrated in nonhuman primates and laboratory animals. Therefore, the clinical data as well as animal data suggest that ethanol consumption is a positive risk factor for hyperprolactinemia. In animal studies, it was found that chronic ethanol administration not only elevates plasma levels of PRL but also increases proliferation of pituitary lactotropes. Ethanol action on lactotropes involves crosstalk with estradiol-responsive signaling cascade or estradiol-regulated cell-cell communication. Additionally, it involves suppression of dopamine D2 receptors inhibition of G proteins and intracellular cyclic adenosine monophosphate (cAMP), modulation of transforming growth factor-beta (TGF-β) isoforms and their receptors (TβRII), as well as factors secondary to TGF-β actions, including production of beta-fibroblast growth factor (bFGF) from follicular-stellate cells. The downstream signaling that governs b-FGF production and secretion involves activation of the MAP kinase p44/42-dependent pathway. A coordinated suppression of D2 receptor- and TβRII receptor-mediated signaling as well as enhancement of bFGF activity might be critical for ethanol action on PRL production and cell proliferation in lactotropes.

Copyright © 2010 S. Karger AG, Basel

Hyperprolactinemia in Humans and Animals

Hyperprolactinemia is a condition in which plasma prolactin (PRL) levels are elevated above normal. Hyperprolactinemia, with elevation of serum PRL of more than 200 ng/ml, is characteristically associated with prolactinomas [1]. Hyperprolactinemia causes reproductive dysfunction such as amenorrhea, galactorrhea and infertility in women [2]. Amenorrhea and galactorrhea may occur alone or together [3]. Up to 25% of patients with secondary amenorrhea have been diagnosed with hyperprolactinemia. Many of these patients showed micro-PRL adenomas or macro-PRL adenomas in the pituitary. Men with hyperprolactinemia often exhibit gynecomastia, impotence, low libido and reduced reproductive hormone levels [4].

Alcohol Abuse and Hyperprolactinemia

There are several reports showing evidence for the existence of high levels of PRL in chronic alcoholic men and women [5, 6]. In a study conducted by European scientists, persistent hyperprolactinemia was observed in 16 alcoholic women during a 6-week treatment trial [7]. These patients reported daily alcohol intake of 170 g for a 2- to 16-year period but had no clinical evidence of alcoholic liver cirrhosis. In a study reported by Japanese scientists, 22 of 23 women admitted for alcoholism treatment had PRL levels above normal and ranged between 27 and 184 ng/ml. These women reported drinking an average of 84.1 g of alcohol each day for at least 7 years. None of these patients showed liver cirrhosis, but 10 had hepatitis and the rest had fatty liver [5, 6]. Studies conducted in a Massachusetts hospital reported hyperprolactinemia (22–87 ng/ml) in 6 of 12 alcohol-dependent women who had a history of drinking 75–247 g of alcohol per day for a minimum period of 7 years [6]. Alcohol-induced hyperprolactinemia is also reported in healthy, well-nourished women during residence on a clinical research ward for 35 days [8]. Sixty percent of women in the heavy drinker category (blood alcohol level 109–199 mg/dl) and 50% of moderate drinkers (blood alcohol level 48–87 mg/dl) showed elevated PRL levels, and many of these drinkers had elevated PRL several days after cessation of drinking. Alcohol-induced hyperprolactinemia was also evident in 66 postmenopausal women [9]. The increase in PRL levels in these patients, however, was associated with increased androgen conversion to estradiol, possibly due to liver cirrhosis.

Alcoholic men also showed elevated plasma levels of PRL [10–12]. Male alcoholic patients frequently show evidence of feminization that is manifested by gynecomastia, spider angiomata, palmar erythema and changes in body hair patterns (see review by Van Thiel [4]). Several studies now indicate a potential role for PRL and estradiol in the pathogenesis for the observed feminization [4]. Alcoholic men show a positive association between the presence of clinically apparent gynecomastia and elevated circulating levels of PRL. These patients also show an elevation of plasma levels of estradiol, which is believed to be due to peripheral conversion of weak adrenal androgens to E2s. The gynecomastia found in alcoholic patients is characterized by a proliferation of the stroma and ducts that are known to be estradiol-responsive. PRL also may act synergistically with estradiol and adrenal steroids and may, therefore, contribute to enhanced breast hypertrophy in alcoholic men. Thus, it appears that chronic alcohol administration in humans causes increased estradiol production and PRL elevation.

Alcohol-induced hyperprolactinemia has also been demonstrated in nonhuman primates [13–15] and laboratory animals [16]. Studies conducted in macaque female monkeys showed that in some but not all of the monkeys, the PRL levels were elevated after chronic self-administration of high doses of alcohol (3.4 g/kg/day) [13, 14]. Interestingly, in one of these monkeys, immunocytochemical examination of the pituitary gland showed apparent pituitary hyperplasia [15]. We have recently obtained data showing that ethanol increases plasma PRL levels and pituitary weight in cyclic female rats and ovariectomized rats as well as potentiates estradiol mitogenic effects in ovariectomized female rats [16]. Therefore, the clinical data as well as animal data suggest that ethanol consumption is a positive risk factor for prolactinomas and hyperprolactinemia.

Mechanism of Alcohol Action on Lactotropes

Dopamine D2 Receptor Splicing

Dopamine, a neurotransmitter of hypothalamic origin, has long been known to be a physiological inhibitor of PRL secretion from the lactotropes [17]. Dopamine's inhibitory action of PRL is mediated by dopamine D2 receptor that belongs to pertussis toxin-sensetive Gi/Go protein-coupled receptor family [18]. Earlier studies have provided some indirect evidence for an effect of alcohol on dopaminergic neurotransmission [19]. Dopamine D2 receptors in the brain are decreased in alcoholic patients [20, 21].

The dopamine D2 receptor exists as two alternatively spliced isoforms, short (D2S) and long (D2L), which has an insertion of 29 amino acids in the third intercellular loop. Both isoforms of dopamine D2 receptor are expressed in lactotropes [18]. It has been reported that the D2L receptor displayed lower affinity than the D2S receptor [22]. In addition, each receptor isoform individually couples to a specific Gα protein because the third intercellular loop that has a deletion in D2S seems to play a central role in G protein coupling [23, 24], and the D2S receptor-specific signaling pathway is also reported [25]. Hence, each isoform of the dopamine D2 receptor may have its specific physiological function.

We have determined the effects of ethanol in the presence and absence of estradiol on dopamine D2 receptor mRNA splicing and the PRL-inhibitory response of a dopaminergic agent bromocriptine in the pituitary of Fischer-344 rats and in primary cultures of anterior pituitary cells [26]. It was observed that ethanol alters the ratio of D2 receptor alternative splicing in the normal rat anterior pituitary gland and in primary cultured pituitary cells. In addition, ethanol diminished bromocriptine's inhibition of PRL secretion in primary cultured pituitary cells. A clinical study has been reported in which dopamine-induced PRL decrement was significantly smaller in alcoholics than in controls, although the PRL response to TRH was similar in those groups [27]. Furthermore, the plasma PRL level in alcoholic patients was elevated by the administration of haloperidol (a dopamine antagonist) but was significantly lower than in the haloperidol-administered control group. Interestingly, low responsiveness to dopamine in alcoholic patients was recovered after alcohol detoxification using diazepam and a vitamin $B_1/B_6/B_{12}$ combination [28]. These clinical reports are consistent with the animal data that ethanol reduces bromocriptine's ability to reduce PRL secretion

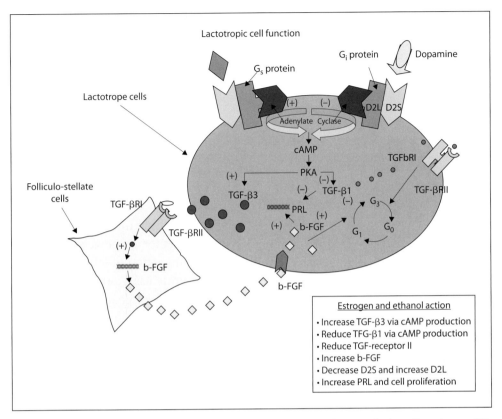

Fig. 1. Diagram summarizing the postulated role of TGF-β in mediation of ethanol's action on PRL production and lactotropic cell proliferation in lactotropes. It is hypothesized that ethanol and estrogen act similarly on lactotropes. Both of these agents decreases dopamine D2 receptor's splice variant D2S to inhibit Gi protein to increase intracellular levels of cAMP, which reduces the function of TGF-β1 signaling via inhibition of TβRII and TβRI dimerization. The increase in intracellular cAMP level also elevates TGF-β3 secretion which enhances the secretion of b-FGF from FS cells. The elevated bFGF is believed to be involved in increased PRL production and enhancement of cell proliferation. The TGF-β-mediated signaling may act as an additional mechanism to that of cAMP and estradiol direct action on PRL gene transcription. + = Stimulation; – = inhibition.

[26]. Furthermore, animal data showed that ethanol alters the D2 receptor splicing and increased the D2L:D2S ratio in pituitary cells in culture and in vivo in the pituitary gland [26]. The D2L receptor displays lower affinity than the D2S receptor [29]. Hence, it could be hypothesized that the reduced dopamine action following ethanol might be partially due to alteration in the expression of D2 receptor isoforms.

It has been reported that estradiol treatment increases the D2L:D2S ratio and reduces dopamine's inhibitory action on PRL secretion in a lactotropic tumor cell line, MMQ cells [30, 31] and in lactotropic cells [26]. These data indicate that there are some similarities between ethanol's and estradiol's actions in the anterior pituitary gland. In this context, it is interesting to note ethanol causes a dose-dependent increase in the transcriptional activity of the ligand-bound, but not the nonligand-bound, estrogen receptor-α level in MCF-7 human breast cancer cells [32]. Estradiol regulates the growth and differentiation of both mammary epithelial cells and lactotropes. Hence, the possibility arises that ethanol may interact with estrogen via altering estrogen receptor levels to regulate lactotropes' function (fig. 1).

G Protein Activation

The D2 receptor is a 7-transmembrane segment protein with a long third intracellular loop and a short intracellular C-terminus. The sixth exon of the D2 receptor gene is often excluded in the mature transcript, resulting in a short (29 amino acids shorter) isoform (D2S). Since the dopamine receptor is the G protein-coupled receptor, the possibility arises that the downstream action of ethanol or estradiol may involve alteration in expression of these G proteins. G proteins belong to the super family of guanine triphosphate (GTP)-binding proteins and function as transducers of information across the cell membrane by coupling diverse receptors to intracellular effectors [33, 34]. All the G proteins are heterotrimers, consisting of α, β and γ subunits. The α subunit has been shown to play a critical role in the regulation of certain intracellular effectors. To date, approximately 20 mammalian α subunit genes have been identified and divided into four major subfamilies (Gs, Gi, Gq and G12) based on the degree of amino acid identity [35]. Gs and Gq mediate stimulatory processes, such as hormone secretion, by activating adenylyl cyclases and phospholipase C, respectively. The members of the Gi subfamily including Gi1, Gi2 and Gi3 generally transduce inhibitory signals by reducing adenylyl cyclase activity and intracellular calcium (Ca^{2+}) levels via potassium (K^+) channel activation and Ca^{2+} channel blockade [36]. The Gi subfamily Gi2 and Gi3 are known to be involved in various cellular processes ranging from cytoskeletal changes to cell growth and oncogenesis [37]. Abnormal expression of G proteins in different target tissues has been described in numerous pathophysiological states including human pituitary adenomas [38]. It is possible for specific types or amounts of G protein available for receptor activation may be relevant to stimulatory or inhibitory effects of hormone or drugs.

Both estradiol and ethanol have been shown to reduce pituitary levels of Gi2 and Gi3 but not Gi1 [39]. The changes in Gi2 and Gi3 protein levels have been shown to be negatively associated with the changes in pituitary weight and PRL content following estradiol or ethanol treatment. These data suggest that both estradiol and ethanol action may depend on inhibition of the function of a negative regulator of lactotropes.

As discussed earlier, both estradiol and ethanol have been shown to increase D2L expression but reduce D2S production [26, 30]. Differences in G protein selectivity of D2S and D2L receptors have been reported previously. Gi2 has been shown to be required for D2L-mediated inhibition of forskolin-stimulated adenylyl cyclase [24], but not D2S-mediated inhibition of Gs-stimulated adenylyl cyclase [40]. Furthermore, Gi3 and Go, but not Gi2, appear to participate in the D2S inhibition of TRH-induced MAPK [23]. It is suggested that the D2S receptor may use Go and Gi3 to inhibit MAPK [23]. Hence, it could be concluded that the mechanisms that mediate estrogen or ethanol action on lactotropic cell growth and prolactin production may involve reduction in the expression of D2S receptor and suppression of the function of inhibitory Gi3 protein.

Role of Growth Factors

Dopamine action on lactotropes is not restricted to PRL gene expression and secretion, as the neurotransmitter also affects lactotrope proliferation. Treatment with dopamine agonists in some patients reverses the hyperprolactinemia [41]. This reversal supports the notion developed from experimental models that prolactinoma formation results from a disruption in dopamine function [42, 43]. We have recently identified a novel signaling cascade for lactotrope growth control involving the D2 receptor short isoform, TGF-β1 and its type II receptor (TβRII) [44]. We found that the tumor promoter estradiol disrupts the signaling cascade and causes loss of growth inhibition by dopamine and TGF-β1 in lactotropes. Expression of D2S in tumor cells induces TGF-β1 and TβRII, recovers dopamine- and TGF-β1-induced growth inhibition and reduces tumor cell growth rate and pituitary tumor transforming gene expression. These data suggest that TGF-β1–mediated D2 receptor signaling prevents the growth of prolactinomas. We found that TGF-β1 levels were elevated by dopamine. Also, TGF-β1 levels have been shown to be decreased after estradiol treatment [45] and ethanol treatment [46]. Thus, by inhibiting dopamine secretion and thereby inhibiting D2 receptor activation, ethanol and estradiol could inhibit TGF-β1 expression and release. It has been shown that D2S overexpression enhances TGF-β1 production and release as well as increases TβRII receptor expression [44]. Hence, ethanol and estradiol could also inhibit TGF-β1 production by reducing D2S receptor expression. The inhibitory effect of dopamine on PRL release of lactotropes is mediated by D2 receptors and involves calcium as well as G protein coupling, inhibition of adenyl cyclase, reduction of intercellular cyclic adenosine monophosphate (cAMP) and a decrease in PRL gene transcription [17]. Although the sequence of TGF-β1 promoter contains no element obviously similar to the cAMP-responsive element core sequence, the TGF-β1 promoter contains several AP-2–like sequence elements that could potentially mediate the cAMP response [47]. Additionally, cAMP analogues inhibit TGF-β1 gene transcription in the pituitary [48]. Thus, it is possible that cAMP-dependent mechanisms may be involved in dopamine regulation of TGF-β1. It could be hypothesized that ethanol action on the signaling cascade for lactotrope growth control involves reduction of D2S, TGF-β1 and TβRII expression. Hence, there is a possibility that ethanol-increased lactotrope proliferation, like estradiol-induced proliferation, might involve the D2S–TGF-β1–TβRII signaling cascade.

Role of Folliculo-Stellate Cells and Lactotropes Communication

Ethanol stimulates lactotropic cell proliferation in primary cultures of mixed pituitary cells but not enriched lactotropes [49]. These data suggest that cell-cell communication between lactotropes and other neighboring cells are necessary for ethanol's mitogenic action. We have found similar cell-cell communication dependency in estradiol's mitogenic action. We have found that both estradiol and ethanol increase TGF-β3 production and secretion from lactotropes [45, 46]. TGF-β3 is transported to FS (FS) cells to cause the release of bFGF, which act on lactotropes to activate the cell-cycle gene for lactotrope proliferation [45].

FS cells are agranular AP cells that extend long processes into areas of glandular endocrine cells and form follicular lumina [50]. These cells participate in phagocytic activity as scavenger cells, play a role in ion transport regulation and are important in metabolism [51]. Traditionally, FS cells were thought to carry out only these supportive functions. It is now known that these cells may have a wide variety of functions and may interact with the other AP cell types as regulatory cells. FS cells have been shown to produce bFGF [52], which increases lactotropic proliferation and is expressed in response to estradiol [45].

bFGF appears to be involved in TGF-β3-stimulated lactotropic proliferation. bFGF is secreted by FS cells under the influence of TGF-β3, and also stimulates lactotropic proliferation. Members of the FGF family of peptides may also have estradiol-potentiated effects on cell status. Estrogen increases the number of FGF receptors in the pituitary cells [53]. In a recent study, we determined the interactive effects of ethanol and TGF-β3 on basic fibroblast growth factor (bFGF) release from folliculostellate (FS) cells and the role of the mitogen-activated protein kinase (MAPK) pathway in this interaction [54]. We found that TGF-β3 and ethanol alone increased release of bFGF from FS cells, but together they showed markedly increased levels of bFGF compared with the individual effect. Ethanol and TGF-β3 alone moderately increased activation of MAPK p44/42, but together they produced marked activation of MAPK p44/42. TGF-β3 alone increased the activation of smad2. Ethanol did not activate smad2 or alter TGF-β3 activation of smad2. Pretreatment of FS cells with a MAPK 1/2 inhibitor or with a protein kinase C (PKC) inhibitor suppressed the TGF-β3 and ethanol actions on MAPK p44/42 activation and bFGF release. Ethanol and TGF-β3, either alone or in combination, increased the levels of active Ras. Furthermore, the MAPK p44/42 activation by TGF-β3 and ethanol was blocked by overexpression of Ras N17, a dominant negative mutant of Ras p21. These data suggest that the PKC-activated Ras-dependent MAPK p44/42 pathway is involved in the crosstalk between TGF-β3 and ethanol to increase bFGF release from FS cells.

Like ethanol, we found involvement of the PKC-dependent p44/42 MAP kinase signaling pathway for crosstalk between estradiol and TGF-β3 in increasing bFGF in folliculostellate cells [55].

Conclusions

Excessive alcohol intake for a long time causes several reproductive abnormalities in humans [4, 6]. The effect of alcohol in men causes hypogonadism, which is manifested by reduced testosterone, testicular atrophy, infertility and loss of libido; in women, it causes anovulation, which is associated with chronic hyperprolactinemia and abnormal secretion of gonadal steroid hormones. We found that chronic ethanol administration not only elevates plasma levels of PRL but also increases proliferation of pituitary lactotropes. In addition to this novel cell-proliferative action of ethanol, we found that ethanol affects estradiol-responsive hormones and polypeptide expression that promotes the growth and transformation of lactotropes. The data presented in this review indicate that ethanol exposure may lead to hyperprolactinemia via orchestrated events involving dopamine D2 receptors, Gi3 protein, TGF-β isoforms and their receptors, as well as factors secondary to TGF-β action. These data demonstrate that ethanol, like estradiol, suppresses dopamine D2 receptor's splice variant D2S to reduce Gi3 protein and activate cAMP signaling to increase PRL production and secretion. Ethanol also reduces TGF-β1-TβRII inhibitory signal on lactotropic growth and PRL synthesis as well as increasing TGF-β3 production and secretion from lactotropes. The secreted TGF-β3 is transported to the neighboring FS cells, where it acts to induce the release of bFGF. FS cell-derived b-FGF stimulates lactotropic cell proliferation as lactotropes escape from TGF-β1 growth inhibition.

Acknowledgement

The research work was supported by National Institutes of Health grant AA 11591.

References

1 Kelley WN (ed): Text Book of Internal Medicine. New York, Lippincott, 1996, pp 1–1983.
2 MacLeod RM, Scapagnini U, Thorner MO: Pituitary Microadenomas. New York, Academic Press, 1980, pp 683–855.
3 Tonner D, Schlechte J: Contemporary therapy of prolactin-secreting adenomas. Am J Med Sci 1993; 306:395–397.
4 Van Thiel DH: Feminization of chronic alcoholic men: a formulation. Yale J Biol Med 1979;52:219–225.
5 Seki M, Yoshida K, Okamura Y: A study on hyperprolactinemia in female patients with alcoholics. Jpn J Alcohol Drug Depend 1991;26:49–59.
6 Teoh SK, Lex BW, Mendelson JH, Mello NK, Cochin J: Hyperprolactinemia and macrocytosis in women with alcohol and polysubstance dependence. J Stud Alcohol 1992;53:176–182.

7 Välimäki M, Pelkonen R, Härkönen M, Tuomala P, Koistinen P, Roine R, Ylikahri R: Pituitary-gonadal hormones and adrenal androgens in non-cirrhotic female alcoholics after cessation of alcohol intake. Eur J Clin Invest 1990;20:177–181.

8 Mendelson JH, Mello NK: Chronic alcohol effects on anterior pituitary and ovarian hormones in healthy women. J Pharmacol Exp Ther 1988;245: 407–412.

9 Gavaler JS: 1994 Aging and alcohol: the hormonal status of postmenopausal women; in Sarkar DK, Barnes C (eds): Reproductive Neuroendocrinology of Aging and Drug Abuse. Boca Raton, CRC Press, 1994, pp 365–378.

10 Ida Y, Isujimararu S, Nakamaura K, Shirrao I, Mukasa H, Egami H, Nakazawa Y: Effects of acute and repeated alcohol ingestion on hypothalamic-pituitary-gonadal and hypothalamic-pituitary-adrenal functioning in normal males. Drug Alcohol Depend 1992;31:57–64.

11 Marchesi C, De Risio C, Campanini G, Majgini C, Piazza P, Grassi M, Chiodera P, Criro V: TRH test in alcoholics: relationship of the endocrine results with neuroendocrinological and neuropsychological findings. Alcohol Alcohol 1992;27:531–537.

12 Soyka M, Gorig E, Naber D: Serum prolactin increase induced by ethanol-a dose dependent effect not related to stress. Psychoneuroendocrinology 1991;16:441–446.

13 Kornet M, Goosen C, Thyssen JH, Van Ree JM: Endocrine profile during acquisition of free-choice alcohol drinking in rhesus monkeys: treatment with desglycinamide-(Arg8)-vasopressin. Alcohol Alcohol 1991;27:403–410.

14 Mello NK, Mendelson JH, Bree MP, Skupny A: Alcohol effects on naloxone-stimulated luteinizing hormone, follicle-stimulating hormone and prolactin plasma levels in female Rhesus monkeys. J Pharmacol Exp Ther 1988;245:895–904.

15 Mello NK, Bree MP, Mendelson JH, Ellingboe J, King NW, Sehgal P: Alcohol self-administration disrupts reproductive function in female macaque monkeys. Science 1983;221:677–679.

16 De A, Boyadjieva N, Pastorcic M, Sarkar DK: Potentiation of estrogen's mitogenic effect on the pituitary gland by alcohol consumption. Int J Oncol 1995;7:643–648.

17 Ben-Jonathan N, Hnasko R: Dopamine as a prolactin (PRL) inhibitor. Endocr Rev 2001;22:724–763.

18 Missale C, Nash SR, Robinson SW, Jaber M, Caron MG: Dopamine receptors: from structure to function. Physiol Rev 1998;78:189–225.

19 Robbins TW, Everitt BJ: Drug addiction: bad habits add up. Nature 1999;398:567–570.

20 Tupala E, Hall H, Bergstrom K, Sarkioja T, Rasanen P, Mantere T, Callaway J, Hiltunen J, Tiihonen J: Dopamine D-2/D-3-receptor and transporter densities in nucleus accumbens and amygdala of type 1 and 2 alcoholics. Mol Psychiatry 2001;6:261–267.

21 Blum K, Noble EP, Sheridan PJ, Finley O, Montgomery A, Ritchie T, Ozkaragoz T, Fitch RJ, Sadlack F, Sheffield D: Association of the A1 allele of the D2 dopamine receptor gene with severe alcoholism. Alcohol 1991;8:409–416.

22 Dal Toso R, Sommer B, Ewert M, Herb A, Pritchett DB, Bach A, Shivers BD, Seeburg PH: The dopamine D2 receptor: two molecular forms generated by alternative splicing. EMBO J 1989;8:4025–4034.

23 Albert PR: G protein preferences for dopamine D2 inhibition of prolactin secretion and DNA synthesis in GH(4) pituitary cells. Mol Endocrinol 2002;16: 1903–1911.

24 Senogles SE: The D2 dopamine receptor isoforms signal through distinct Gi alpha proteins to inhibit adenylyl cyclase: a study with site-directed mutant Gi alpha proteins. J Biol Chem 1994;269:23120–23127.

25 Senogles SE: The D2s dopamine receptor stimulates phospholipase D activity: a novel signaling pathway for dopamine. Mol Pharmacol 2000;58:455–462.

26 Oomizu S, Boyadjieva N, Sarkar DK: Ethanol and estradiol modulate alternative splicing of dopamine D2 receptor mRNA and abolish the inhibitory action of bromocriptine on prolactin release from the pituitary gland. Alcohol: Clin Exp Res 2003;27: 975–980.

27 Marchesi C, Ampollini P, Chiodera P, Volpi R, Coiro V: Alteration in dopaminergic function in abstinent alcoholics. Neuropsychobiology 1997;36:1–4.

28 Markianos M, Moussas G, Lykouras L, Hatzimanolis J: Dopamine receptor responsivity in alcoholic patients before and after detoxification. Drug Alcohol Depend 2000;57:261–265.

29 Dal Toso R, Sommer B, Ewert M, Herb A, Pritchett DB, Bach A, Shivers BD, Seeburg PH: The dopamine D2 receptor: two molecular forms generated by alternative splicing. EMBO J 1989;8:4025–4034.

30 Guivarc'h D, Vincent JD, Vernier P: Alternative splicing of the D2 dopamine receptor messenger ribonucleic acid is modulated by activated sex steroid receptors in the MMQ prolactin cell line. Endocrinology 1998;139:4213–4221.

31 Livingstone JD, Lerant A, Freeman ME: Ovarian steroids modulate responsiveness to dopamine and expression of G-proteins in lactotropes. Neuroendocrinology 1998;68:172–179.

32 Fan S, Meng Q, Gao B, Grossman J, Yadegari M, Goldberg ID, Rosen EM: Alcohol stimulates estrogen receptor signaling in human breast cancer cell lines. Cancer Res 2000;60L:5635–5639.

33 Gilman AG: G proteins: transducers of receptor-generated signals. Ann Rev Biochem 1987;56:615–649.
34 Simon MI, Strathmann MP, Gautam N: Diversity of G proteins in signal transduction. Science 1991;252:802–808.
35 Wilkie TM, Gilbert DJ, Olsen AS, Chen XN, Amatruda TT, Korenberg JR, Trask BJ, de Jong P, Reed RR, Simon MI, Jenkins NA, Copeland NG: Evolution of the mammalian G protein alpha subunit multigene family. Nat Genet 1992;2:85–91.
36 Vallar L, Meldolesi J: Mechanisms of signal transduction at the dopamine D2 receptor. Trends Pharmacol Sci 1989;10:74–77.
37 Dhanasekaran N, Dermott JM: Signaling by the G12 class of G proteins. Cell Signal 1996;8:235–245.
38 Ballare E, Mantovani S, Bassetti M, Lania A, Spada A: Immunodetection of G proteins in human pituitary adenomas: evidence for a low expression of proteins of the Gi subfamily. Euro. J Endocrinology 1997;137:482–489.
39 Chaturvedi K, Sarkar DK: Alteration in G proteins and prolactin levels in pituitary after ethanol and estrogen treatment. Alcohol Clin Exp Res 2008;32:806–813.
40 Ghahremani MH, Cheng P, Lembo PM, Albert PR: Distinct roles for Gαi2, Gαi3, and Gβγγin modulation of forskolin- or Gs-mediated cAMP accumulation and calcium mobilization by dopamine D2S receptors. J Biol Chem 1999;274:9238–9245.
41 Tonner D, Schlechte J: Contemporary therapy of prolactin-secreting adenomas. Am J Med Sci 1993;306:395–397.
42 Asa SL, Kelly MA, Grandy DK, Low ML: Pituitary lactotroph adenomas develop after prolonged lactotroph hyperplasia in dopamine D2 receptor-deficient mice. Endocrinology 1999;140:5348–5355.
43 Sarkar DK, Gottschall PE, Meites J: Damage to hypothalamic dopaminergic neurons is associated with development of prolactin-secreting tumors. Science 1982;218:684–686.
44 Sarkar DK, Chaturvedi K, Oomizu, S, Boyadjieva N, Chen CP: Dopamine and TGF-β1 interact to inhibit the growth of pituitary lactotropes. Endocrinology 2005;146:4179–4188.
45 Hentges S, Sarkar DK: Transforming growth factor β regulation of estradiol-induced prolactinomas. Front Neuroendocrinol 2001;22:340–363.
46 Sarkar DK, Boyadjieva NI: Ethanol alters production and secretion of estrogen-regulated growth factors that control prolactin-secreting tumors in the pituitary. Alcohol Clin Exp Res 2007;31:2101–2105.
47 Geiser AG, Kim SJ, Roberts AB, Sporn MB: Characterization of the mouse transforming growth factor-beta 1 promoter and activation by the Ha-ras oncogene. Mol Cell Biol. 1991;11:84–92.
48 Pastorcic M, Sarkar DK: Down regulation of TGF-β1 gene expression in anterior pituitary cells treated with forskolin. Cytokine 1997;9:106–111.
49 De A, Boyadjieva N, Oomizu S, Sarkar DK: Ethanol induces hyperprolactinemia by increasing prolactin release and lactotrope growth in female rats. Alcoholism: Clinical and Experimental Research 2002; 26:1420–1429.
50 Vila-Porcile E, Olivier L: The problem of folliculo-stellate cells in the pituitary gland; in Motta PM (ed): Ultrastructure of Endocrine Cells and Tissue. Boston, Martinus Nijhoff, 1984, pp 64–71.
51 Allaerts W, Denef C: Regulatory activity and topological distribution of folliculo-stellate cells in rat anterior pituitary cell aggregates. Neuroendocrinology 1989;49:409–418.
52 Ferrara N, Schweigerer L, Neufeld G, Mitchell R, Gospodarowicz D: Pituitary follicular cells produce basic fibroblast growth factor. Proc Natl Acad Sci USA 1987;84:5773–5777.
53 Takahashi H, Nakagawa S: Effects of estrogen and fibroblast growth factor receptor induction in MtT/Se cells. Endocrine Res 1997;23:95–104.
54 Chaturvedi K, Sarkar DK: Role of PKC-Ras-MAPK p44/42 in ethanol and TGF-β3-induced bFGF release from folliculostellate cells. J Pharm Exp Ther 2005;314:1346–1352.
55 Chaturvedi K, Sarkar DK: Involvement of PKC dependent p44/42 MAP kinase signaling pathway for cross-talk between estradiol and TGF-β3 in increasing bFGF in folliculostellate cells. Endocrinology 2004;145:706–715.

Dipak K. Sarkar
Endocrine Program, Biomedical Division of the Center of Alcohol Studies and Department of Animal Sciences
Rutgers, The State University of New Jersey, 67 Poultry Farm Lane
New Brunswick, NJ 08901–8525 (USA)
Tel. +1 732 932 1529, Fax +1 732 932 4134, E-Mail sarkar@AESOP.rutgers.edu

Experience from the Argentine Pegvisomant Observational Study: Preliminary Data

N. García Basavilbaso[a,b] · M. Guitelman[a,b] · A. Nagelberg[a] · G. Stalldecker[c] · A. Carabelli[c] · O. Bruno[d] · K. Danilowitz[d] · M. Manavela[d] · S. Mallea Gil[e] · C. Ballarino[e] · R. Guelman[f] · D. Katz[g] · S. Fidalgo[h] · R. Leal[h] · H. Fideleff[i] · M. Servidio[i] · D. Bruera[l] · F. Librandi[j] · A. Chervin[k] · M. Vitale[k] · A. Basso[b]

[a]Hospital Carlos G. Durand, [b]INBA Instituto de Neurocirugía de Buenos Aires, [c]Hospital Pirovano, [d]Hospital de Clínicas, [e]Hospital Militar, [f]Htal. Italiano, [g]Fleni, [h]Hospital Churruca, [i]Hospital Alvarez, [j]Hospital Rivadavia, and [k]Hospital Santa Lucía, Buenos Aires, and [l]Clínica Caraffa, Córdoba, Argentina

Abstract

The GH receptor antagonist pegvisomant is an efficient agent to achieve biochemical control of acromegaly in those cases refractory to surgery and medical therapy with somatostatin analogs. We conducted an observational multicenter study consisting of data collection in accordance with the standard management of patients with acromegaly in everyday practice. We reviewed the medical records of 28 patients, 23 females, who were treated with pegvisomant due to the lack of biochemical response or intolerance to the somatostatin analogs. The objective was to monitor long-term safety and efficacy of the antagonist. 82% of the patients had previous pituitary surgery, 53.6% radiotherapy and 96.4% received medical therapy for acromegaly. Only 19.2% of the patients had pituitary residual tumor size larger than 1 cm, the remainder harbored a microadenoma or no visible tumor in the pituitary images. In terms of biochemical efficacy, IGF-I levels decreased to normal ranges in 45% and 58.8% of patients after 3 and 6 months of treatment, respectively, the daily mean dose of pegvisomant being 9.6 ± 1.1 mg. Adverse events, potentially related to pegvisomant were reported in 6 patients (21.4%), local injection site reaction and elevated liver enzymes being the most frequent. Tumor size did not show enlargement in the evaluated population (15 patients) during the period of the study. This paper presents preliminary data from a small observational study in Argentina which represents the first database in our country.

Copyright © 2010 S. Karger AG, Basel

Medical therapy of acromegaly has undergone remarkable advances over the last two decades. Historically, therapeutic choices were limited to surgery or irradiation. Currently, several classes of pharmacologic agents are available for use as secondary, and in some instances, primary therapy. The three categories of available agents

Table 1. Long-term studies of pegvisomant monotherapy

	Study design	Number of patients	% of patients with normal IGF-I	Pegvisomant mean dose mg QD	Study duration months
Trainer et al. [1]	prospective, double-blind, placebo-controlled	26 26 28	54 81 89	10 15 20	3 3 3
Van der Lely et al. [2]	prospective, uncontrolled	152	97	19.6	18
Colao et. al. [4]	prospective uncontrolled	16	75	24.7	12
Schreiber et al. [3]	multicenter, observational	177	64 76.3	16.5	6 24

include: dopamine agonists, somatostatin analogs and growth hormone (GH) receptor antagonists.

Pegvisomant, a genetically modified GH analog, acts as a GH receptor antagonist and suppresses IGF-I levels. The structure of pegvisomant is similar to the native GH with the exception of 9 amino acid substitutions. These targeted substitutions permit the hormone to bind to the GH receptor without receptor activation. Thus, pegvisomant functions as a competitive antagonist, preventing normal endogenous GH from binding to its receptor transduction and IGF-I synthesis and secretion. On the other hand, pegylation of the analogue reduces its clearance rate, increases circulating half-life, possibly reducing antigenicity, and reduces the affinity for GHR, which means that, compared with GH, up to a 1,000-fold higher concentration of pegvisomant is needed in order to achieve adequate GH antagonism. Pegvisomant is an efficient agent to achieve biochemical control of acromegaly in those cases refractory to surgery and medical therapy with somatostatin analogs.

Table 1 shows three prospective studies and one observational study on the long-term efficacy of pegvisomant monotherapy.

The safety profile of pegvisomant is favorable, but strict follow up to exclude significant hepatotoxic effects and tumor growth is mandatory.

Objective

We conducted an observational multicentric study consisting of data collection in accordance with the standard management of patients with acromegaly in everyday practice and with the recommendations of the pegvisomant summary of product characteristics.

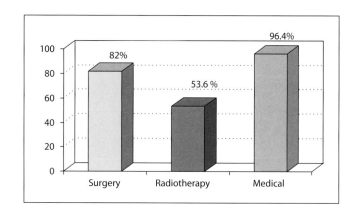

Fig. 1. Previous treatments.

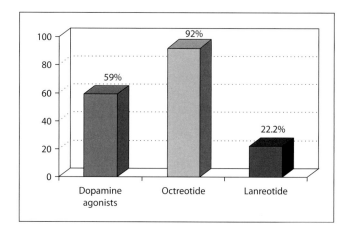

Fig. 2. Previous medical therapy.

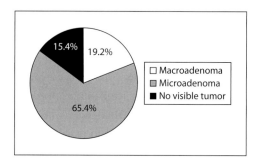

Fig. 3. Tumor residual size before treatment with pegvisomant.

Subjects and Methods

We reviewed the medical records of 28 patients.

Indication for pegvisomant treatment and treatment monitoring were at the discretion of the treating physician's decision. Serum IGF-I levels have been measured in local laboratories and interpreted according to local, age-dependent reference ranges.

Results were given as total IGF-I levels; also, in an attempt to obtain a better comparability, they were also presented as x-fold of the individual upper limit of normal value (ULN).

Results

Baseline Characteristics

The 28 documented patients had an average history of 7.15 years of acromegaly with a mean age at diagnosis of 42 ± 12.6 years.

All but 2 patients had acromegaly due to a GH-secreting pituitary adenoma, 22 macroadenoma and 4 microadenoma.

In 2 patients, autonomous GH hypersecretion was documented clinically and biochemically without morphological evidence of pituitary tumor by magnetic resonance imaging (MRI). No evidence of ectopic disease was found.

Deficiencies of other pituitary hormones were documented at baseline before treatment with pegvisomant: LH/FSH deficiency in 25, TSH deficiency in 14.3, and ACTH deficiency in 10.7%. Hyperprolactinemia was found in 5% of the patients.

Previous Treatments

A total of 28 patients were reviewed: 23 (82%) had previous pituitary surgery, 15 had previous irradiation therapy (53.6%), and 27 (96.4%) had previous medical therapy for acromegaly (fig. 1).

Medical therapy, i.e. dopamine agonists, octreotide, or lanreotide, had been given to 16 (59%), 25 (92%), and 6 (22.2%) patients, respectively (fig. 2).

Tumor Size

The size of pituitary residual adenoma at baseline before pegvisomant was considered comparable to a macroadenoma in 5 of 26 patients (19.2%), and to a microadenoma in 17 patients (65.4%).

Four patients (15.4%) had no visible tumor (fig. 3).

Treatment with Pegvisomant

Initially, the mean dose of pegvisomant was 9.6 ± 1.1 mg q.d. Twenty-six patients received 10 mg q.d.; pegvisomant dose was increased to 30 mg q.d. in one of them, while 2 patients started with 10 mg q.o.d. and then the dose was increased to 10 mg q.d.

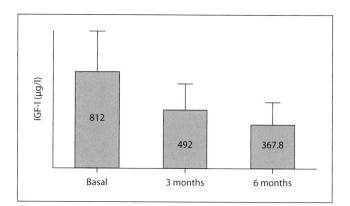

Fig. 4. Course of IGF-1 during treatment with pegvisomant. Mean values ± SD. Baseline to 3 months, p < 0.01; baseline to 6 months, p < 0.001; 3 months to 6 months, n.s.

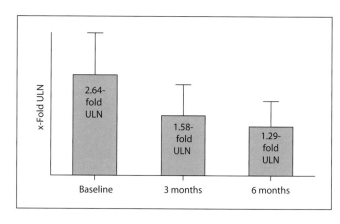

Fig. 5. IGF-1 levels compared with individual reference ranges. Mean values ± SD. Baseline to 3 months, p < 0.01; baseline to 6 months, p < 0.001; 3 months to 6 months, n.s.

Over the first 3 months of treatment, body weight descended from a mean of 86.75 to 86.15 kg (p = 0.42).

IGF-I Levels during Treatment with Pegvisomant

IGF-I was measured and compared in each subject at baseline, and at the 3-month and 6-month visits during treatment with pegvisomant: absolute IGF-I levels decreased from 812.67 ± 341.64 µg/l at baseline to 492 ± 220.97 µg/l at the 3-month visit and 367.8 ± 190.19 µg/l at the 6-month visit (fig. 4).

When compared to individual reference ranges, baseline IGF-I levels decreased from 2.64 ± 1.1-fold ULN to 1.58 ± 0.8-fold ULN at the 3-month visit, and to 1.29 ± 0.67-fold ULN at the 6-month visit (fig. 5).

No patient had normal IGF-I baseline levels. IGF-I levels were within the normal ranges in 9 of 20 patients (45%), and in 10 of 17 patients (58.8%) after 3 and 6 months, respectively (fig. 6).

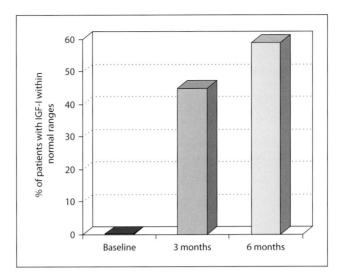

Fig. 6. Percentage of patients that normalized IGF-1 levels under treatment with pegvisomant.

Adverse Events

Nonserious adverse events (AEs) or serious adverse events (SAEs), regardless of their being potentially related to pegvisomant, were reported in 8 patients (29.6%). AEs (local injection site reaction in 2 and elevated liver enzymes in 1) or SAEs (local injection site reaction in 1 and elevated liver enzymes in 2 patients) potentially related to pegvisomant were reported in 6 patients. Nephrotic syndrome and acute myeloblastic leukemia were 2 unrelated adverse events (the former was probably due to a noncontrolled diabetes mellitus). Treatment with pegvisomant was discontinued in 5 patients due to adverse events; in 3 of them it was potentially related to the drug (table 2).

Follow-up MRIs were performed in 15 patients at the 3rd-, 6th-,12th-, 24th- or 36th-month visits; no pituitary tumor size increase was found.

Discussion

The GH receptor antagonist pegvisomant is the newest drug for acromegaly treatment. Its high efficacy shown in clinical studies is particularly relevant for patients with persisting disease activity, despite the use of all other nonmedical and medical treatment options.

Since long-term experience with pegvisomant is limited, this study adds important information on safety and efficacy of pegvisomant for clinical practice.

IGF-1 normalization rates in the present study (58.8%) were lower than those reported in previous clinical studies. This may be explained because almost all patients included in a closely monitored trial have a better compliance, and dose titration of

Table 2. Adverse events

Adverse event	AEs	SAEs	Drug discontinuation	AE related to drug
Local injection site reaction	2	1	1 SAE	yes
Elevated liver enzymes	1	2	2 SAEs	yes
Nephrotic syndrome		1	1	no
Acute myeloblastic leukemia		1	1	no

Pegvisomant follows a pre-assigned dosing algorithm reaching maximum pegvisomant dose.

The three prospective studies follow up a total of 248 patients (Trainer, van der Lely and Colao) with acromegaly for up to 18 months. Normalization of IGF-1 was seen in 54–97% of patients and was dose dependent (10–40 mg q.d.).

Schreiber et al. [3] conducted an observational study in Germany and evaluated the course of IGF-1 during treatment in 177 acromegalic patients. The rate of normalization of 64% at 6 months of treatment also reflects the lack of treatment standardization, similar to that which occurred in our study. After 24 months, the term rate of IGF1 normalization reached 76.3%.

The safety profile of pegvisomant in this study was similar to that shown in clinical studies [1–4]. Elevation of hepatic enzymes to more than three times the upper normal limit was present in 10.7% of our patients. Local injection site reaction was a cause of treatment withdrawal in 1 patient. Tumor size did not show enlargement in the evaluated population (15 patients) during the study period.

This is preliminary data from a small observational study in Argentina that represents the first database in our country.

References

1 Trainer PJ, et al: Treatment of acromegaly with the growth hormone-receptor antagonist pegvisomant. N Engl J Med 2000;342:1171–1177.
2 Van der Lely AJ, Trainer PJ, et al: Long term treatment of acromegaly with pegvisomant, a growth hormone receptor antagonist. Lancet 2001;358:1754–1759.
3 Schreiber I, Strasburger CJ, et al: Treatment of acromegaly with the GH receptor antagonist pegvisomant in clinical practice: safety and efficacy evaluation from the German Pegvisomant Observational Study. Eur J Endocrinol 2007;156:75–82.
4 Colao A, et al: Efficacy of 12-month treatment with the GH receptor antagonist pegvisomant in patients with acromegaly resistant to long-term, high dose somatostatin analog treatment: effect on IGF-1 levels, tumor mass, hypertension and glucose tolerance levels. Eur J Endocrinol 2006;154:467–477.
5 Trainer PJ, et al: Pegvisomant in the treatment of acromegaly. Adv Drug Delivery Rev 2003;55:1303–1314.
6 Parkinson C, Trainer PJ, et al: Gender, body weight, disease activity, and previous radiotherapy influence the response to pegvisomant. J Clin Endocrinol Metab 2007;92:190–195.

7 Frohman LA, Bonert V: Pituitary tumor enlargement in two patients with acromegaly during pegvisomant therapy. Pituitary 2007;10:283–289.
8 Trainer PJ, et al: Disease activity in acromegaly may be assessed 6 weeks after discontinuation of pegvisomant. Eur J Endocrinol 2005;152:47–51.
9 Trainer PJ, et al: Pegvisomant interference in GH assays results in underestimation of GH levels. Eur J Endocrinol 2007;156:315–319.
10 Barkan A, et al: Long term effects of pegvisomant in patients with acromegaly. Nate Clin Pract Endocrinol Metabol 2008;4:324–332.
11 Orskov H, et al: Concomitant, specific determination of growth hormone and pegvisomant in human serum. Growth Horm IGF Res 2007;17:431–434.
12 Colao A, et al: Pegvisomant in acromegaly: why, when, how. J Endocrinol Invest 2007;30:693–699.

Dr. Mirtha Guitelman
División Endocrinología, Hospital Carlos G. Durand
Díaz Velez 5044 (C1405DCS)
Buenos Aires (Argentina)
Tel. +54 11 4982 5212, Fax +54 11 4958 4377, E-Mail mguitelman@speedy.com.ar

Gender Differences in Macroprolactinomas: Study of Clinical Features, Outcome of Patients and Ki-67 Expression in Tumor Tissue

Patricia Fainstein Day[a] · Mariela Glerean[a] · Soledad Lovazzano[a] · Marcelo Pietrani[b] · Silvia Christiansen[c] · Marta Balzaretti[a] · Andrea Kozak[a] · Antonio Carrizo[d]

[a]Endocrine and Nuclear Medicine Unit, [b]Radiology Unit, [c]Pathology Unit, and [d]Neurosurgery Unit, Hospital Italiano, Buenos Aires, Argentina

Abstract

Prolactinomas in men are usually macroprolactinomas and other investigators have attributed bigger size of tumors in men to delay in diagnosis. A retrospective study of 71 macroadenomas (42 men) was carried out. Parameters studied were age, signs and symptoms at presentation, time of onset of symptoms, basal prolactin, estradiol, and total testosterone levels, tumor size and Ki 67 expression in tumor tissue. Male patients were older. Visual defects were significantly more prevalent in men. Hardy 4 stage tumors were found only in men. We found no significant correlation between tumor size and the patients age nor between tumor size and the onset of symptoms. Whereas basal E2 levels (21.2 ± 12.9 vs. 33.3 ± 43.3 pg/ml, p = n.s.) were very similar in male and female patients, testosterone levels were significantly higher in men (0.6 ± 0.5 vs. 1.8 ± 1.2 ng/ml, p = 0.02). The rate of cell proliferation represented by Ki 67 was significantly higher in tumors in men (3.5 ± 1.2 vs. 1.5 ± 0.5%, p = 0.0001). This is the first study focused in macroprolactinomas that shows that they are clinically and biologically more aggressive in men. Hypogonadism in men could appear later in the progression of prolactinomas and this might explain why men were older at the time of diagnosis. Furthermore, testosterone could be a source for E2 in situ aromatization giving male tumors an advantage in cell proliferation.

Copyright © 2010 S. Karger AG, Basel

It is fairly well known that microprolactinomas are more frequently seen in women who consult because of menstrual disturbances and/or galactorrhea. They have an excellent response to dopamine agonists [1–5] and a benign evolution, even without treatment [6]. On the other hand, prolactinomas in men are usually macroprolactinomas and tumor-related symptoms such as visual defects as well as hypogonadism often lead to diagnosis. Since prolactinoma in men is a less-frequent condition, response to dopamine agonists has been studied [2, 7–9] in a significantly lower number of patients as compared to women. In men the natural history of prolactinomas is unknown, probably because the potentially more aggressive course of the disease systematically leads to medical treatment.

It has been suggested that late diagnosis in men could account for gender differences. Hypogonadism symptoms in men are less objective than in women [9, 10]. On the other hand, the nature of prolactinomas in men could be more aggressive. In this case we could expect biological markers of higher proliferation rate [7, 8, 11–14].

Several lines of research support the role of estrogens in prolactin secretion and the development and progression of lactotroph tumors [15–18] and this might have a relationship with the higher prevalence of prolactinomas in women. However, little is known whether there is any influence of testosterone in prolactin secretion and prolactinoma induction. Plasmatic testosterone is systematically screened in men with prolactinomas but basal estradiol and testosterone levels both in male and female patients with prolactinoma have not been studied.

Gender differences were retrospectively studied in our patients with prolactinomas, especially regarding prevalence of micro vs. macroprolactinomas. Furthermore, since most prolactinomas in men are macroprolactinomas, it has been our intention to study the clinical-biochemical differences, including estradiol and testosterone levels in men and women bearing macroprolactinomas and response to a 2-year treatment with cabergoline.

Subjects and Methods

One hundred and forty-seven patients with prolactinoma (105 women), diagnosed between 2000 and 2006 at our Neuroendocrine Department, were included in this study. The criteria to diagnose microprolactinomas were serum prolactin levels ≥100 ng/ml and evidence of pituitary tumor with a maximal diameter ≤10 mm, and for macroprolactinomas, prolactin levels ≥200 ng/ml and evidence on MRI of a pituitary tumor with a maximal diameter >10 mm.

Our study focused on gender differences in 71 macroprolactinomas (40 men). The parameters studied were age, signs and symptoms at presentation, onset of symptoms, basal serum prolactin, estradiol, testosterone and basal tumor size, prolactin levels and tumor size after 24 months' treatment with cabergoline and Ki 67 expression in tumor tissue of patients referred to surgical treatment. Basal prolactin, estradiol and testosterone levels were assessed using commercial chemiluminescent enzyme immunoassay kits in the same laboratory. Analytic sensibility was 0.6 ng/ml for prolactin, 10.0 pg/ml for estradiol and 0.15 ng/ml for testosterone. Tumor size was evaluated by MRI and diameters (mm) were measured by the same radiologist.

Fragments of each pituitary tumor were embedded in paraffin for pathological diagnosis. Prolactin and Ki 67 were assessed by indirect immunoperoxidase method with anti prolactin and Mib 1 antibodies.

Hypogonadism was defined as oligo-amenorrhea in women and serum testosterone levels below 2.8 ng/ml in men. All but one of the women bearing macroprolactinomas had amenorrhea.

Table 1. Clinical features of 71 macroprolactinomas

	Women	Men	p
Patients, %	43.2	56.7	n.s.
Age (mean ± SD) (r = 15–81 years)	34.2±10.1	41.7±14.6	<0.03
Onset of symptoms, years	4.0±3.7 (r: 0.4–12)	2.0±3.8 (r: 0–20)	0.059
Headaches, %	55.6	68.4	n.s.
Hypogonadism %	100	68.5	<0.04
Visual defects %	27.6	42.1	<0.01
Serum prolactin, ng/ml	880.0±1,491.2	4634.0±7,468.7	<0.01
Tumor size, maximum diameter, mm	21.7±12.0	33.6±24.4	0.028

Regarding statistical analysis, data were reported as mean ± SD. Statistical analysis was performed by means of SSPS and statistical significance was set at 5%. Correlations were performed by calculating the Spearman and Pearson coefficients where appropriate.

Results

In our study, 73% of 147 prolactinomas were carried by female patients, 70 percent of tumors in women were microprolactinomas and all but 2 prolactinomas in men were macroprolactinomas.

A contrasting analysis of the features of 71 macroprolactinomas in both genders (40 men) was carried out. Male patients were older. The time elapsed from the onset of symptoms to diagnosis was shorter in men although not significantly. Hypogonadism was more prevalent in females and visual defects were more prevalent in men (table 1).

Prolactin levels were significantly higher in men as well as tumor size (table 1). Prolactin levels were positively correlated with tumor size (r2: 0.68, p = 0.02). Invasive tumors were found only in men (fig. 1). Importantly, we found no significant correlation between tumor size and age of patients (r^2 = 0.08, p = n.s.) nor between tumor size and the onset of symptoms (r = 0.12, p = n.s.) and male patients carried larger tumors no matter their age (not shown).

Basal circulating estradiol levels were very similar in male and female patients and testosterone levels were significantly higher in men (fig. 2). No significant correlation was found between prolactinemia and estrogen/testosterone levels nor between tumor size and estrogen/testosterone levels (not shown).

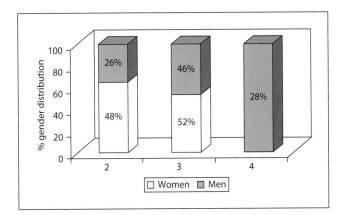

Fig. 1. Tumoral stages (Hardy classification) in 71 macroprolactinomas. Invasive tumors were only seen in male patients.

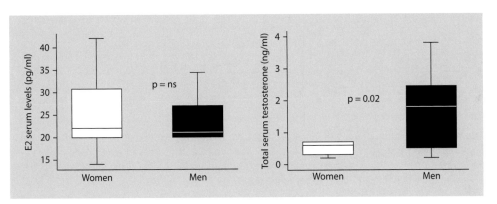

Fig. 2. Basal testosterone and estradiol levels in macroprolactinomas. Circulating estradiol levels were similar in both genders and testosterone levels were higher in male patients.

We found no significant differences between female (n = 28) and male (n = 33) patients in prolactin levels decrease (82.2 ± 6.5 vs. 98.1 ± 8.4%, respectively, p = n.s.) and tumor shrinkage (87.5 ± 49.0 vs. 96.1 ± 55.0%, respectively, p = n.s.) after 2 years' treatment with cabergoline. The cabergoline dose was also similar (1.2 ± 1.6 vs. 1.9 ± 1.5 mg/w, respectively, p = n.s.).

Gillan et al. [19] published a case report of a male patient with a very resistant tumor in whom prolactin levels increased after testosterone supplementation. Treatment with anastrozole, which is an aromatase inhibitor, led to normalization of prolactin levels and made it possible to reduce cabergoline dose. We had a similar case in a 47-year-old patient with a responsive tumor to cabergoline treatment. When testosterone supplementation was added, prolactin levels increased and they fell when testosterone supplementation was withdrawn. During the treatment with anastrozole and testosterone the lowest prolactin levels could be reached even with a lower cabergoline dose (fig. 3).

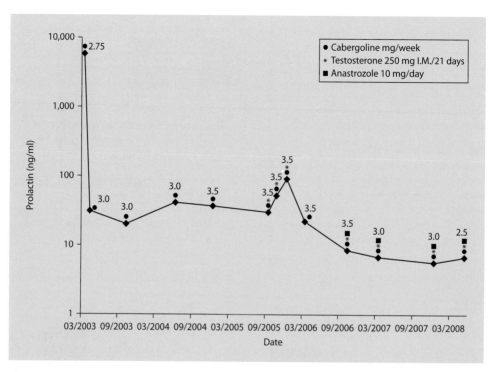

Fig. 3. A 47-year-old patient with macroprolactinoma under cabergoline treatment. Prolactin levels increased when testosterone supplementation was added and decreased when testosterone was withdrawn. After addition of anastrozole to testosterone supplementation the lowest prolactin levels were reached even with lower cabergoline dose.

Among the 18 patients referred to surgical treatment (11 men), 8 presented acute neurologic symptoms, 2 had cystic adenomas (1 man), 1 female patient had cabergoline resistance, in 3 cases surgical treatment was due to the patients' choice (2 women) and in the remaining 4 operated patients, treatment was suggested by the neurosurgeon mainly due to large tumor size (3 women). No patient was cured by surgery, all needing to be kept on cabergoline treatment.

The rate of cell proliferation represented by Ki 67 was significantly higher in men tumors (fig. 4). No significant correlation was found between Ki 67 expression and tumor size nor with testosterone or estrogen levels (not shown).

Discussion

The greater prevalence of prolactinomas in women and the fact that 70% of them were microprolactinomas while all but two of the prolactinomas carried by men were macroprolactinomas, as shown by our 6-year retrospective study, support the gender differences described by other authors [1–4, 20].

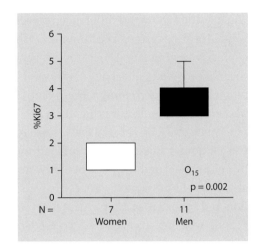

Fig. 4. Gender differences in % Ki-67 expression in 18 macroprolactinomas.

When the comparison was focused on macroprolactinomas, gender differences regarding tumor size were still present. Male patients had larger tumors, invasive tumors were exclusively observed in them and they showed more prevalence of visual defects. Other neurological symptoms as seizures, acute hydrocephaly and comma led to surgical treatment in 8 male patients.

Colao et al. [9] have also studied gender differences focusing on macroprolactinomas. In their study, male patients had larger tumors but without more prevalence or severity of neurological signs. An explanation for this could be that our male patients carried bigger tumors.

Age at presentation of our male patients is in agreement with that reported by other investigators [8–11]. For most researchers [9, 10, 20] the explanation of the prevalence of large tumors in men would account for a delay in diagnosis mainly owing to the lack of perception of symptoms of hypogonadism as compared to females. The fact that men with macroprolactinoma were older in our study would support this hypothesis.

However, we do not share this view completely. Our findings in the whole population of prolactinomas showed that, regardless the age at the time of diagnosis, prolactinomas in men are always larger and, particularly in the case of macroprolactinomas, we did not find any correlation between age at diagnosis and tumor size or length of the symptoms. These findings are in agreement with those of Delgrange et al. [12]. On the other hand, the natural history of prolactinomas in women shows that only a third of them became larger [21].

Interesting enough, 30% of men with macroprolactinomas in our and other studies [20, 22] did not have testosterone deficiency despite its larger tumor size, whereas 100% of women suffered from hypogonadism. The highest resistance of gonadotrophic axis in men has been previously observed [19, 20].

Our study shows no difference regarding the circulating estrogen level in men and women bearing macroprolactinoma. On the other hand, higher testosterone levels in men could have a stimulating effect for prolactin secretion through estrogen aromatization at the pituitary. Kadioglu et al. [23] showed that mRNA aromatase is expressed in normal pituitary in men and women. Haug et al. [16] showed that physiologic levels of testosterone stimulated prolactin production in GH3 cells whereas 5α-dihydrotestosterone did not. Gillan et al. [19] showed that testosterone supplementation in male patients carrying prolactinomas would increase prolactin secretion and that this effect could be restored by anastrozole, and the present study confirms their result. Data on mitotic action of androgens in rat pituitary are controversial [24]. Tumor induction by GH3 implantation and testosterone supplementation was demonstrated in rats [16]. Carretero et al. [25] showed that aromatase is increased in spontaneous prolactinomas in rats, thus suggesting the possibility that an abnormal high conversion of testosterone into estradiol in pituitary cells may contribute to the genesis of prolactinomas.

Ki-67 antigen is a proliferation marker that is selectively expressed during G1, S1, G2 and M phases of the cell cycle [26, 27]. A high Ki-67 proliferative index in a pituitary adenoma might indicate a more aggressive behavior [28, 29]. Wierinckx et al. [30] focused their study on Ki 67 and other markers and genes implicated in invasiveness of 25 prolactinomas and the postsurgical outcome of the patients. They found that Ki 67 expression in more than 2% of the cells suggested a subtype of 'agressive-invasive' prolactinomas. Interestingly, in our study, none of the prolactinomas of female patients and all but one of prolactinomas of the male patients showed Ki 67 expression above 2%.

Although surgery indication related to acute neurological complications was observed only in our male patients, we did not see any gender difference in the outcome of patients treated with cabergoline. Similar results were observed by Colao et al. [9]. In the series of Verlhelst et al. [31], men were reported to have worse results regarding lowering prolactin levels than women treated with dopamine agonists but micro and macroprolactinomas were not analyzed separately in that study. Resistance to dopaminergic agents is related to dopamine receptor type 2 (D2R) bioavailability of prolactinomas [32] and, on the other hand, clinical response to dopamine agonist does not rule out tumors aggressive behavior [33]. It has been shown by others that cabergoline treatment is the first line therapy for micro- and macro- [1–5, 31, 34–39] and also for giant prolactinomas [19, 34]. So, we believe that the fact that men and women had a similar response to cabergoline treatment does not exclude that prolactinomas in men tend to be more aggressive than in female patients.

Concluding Remarks

This is the first study focused on macroprolactinomas that shows that they are clinically and biologically more aggressive in men as compared to women. However, hypogonadism was significantly less prevalent in men given the fact that 30% of the patients had normal testosterone levels regardless higher prolactinemia. Hypogonadism and their symptoms in men could appear later in the progression of prolactinomas and this might explain the fact that male patients were older at the time of diagnosis. Furthermore, testosterone levels could be a source for estradiol in situ biosynthesis giving male tumors an advantage in cell proliferation. Addition of aromatase inhibitors might be an option for a subset of male patients with macroprolactinomas.

References

1 Schlechte JA: Long-term management of prolactinomas. J Clin Endocrinol Metab 2007;92:2861–2865.
2 Melmed S: Update in pituitary disease. J Clin Endocrinol Metab 2008;93:331–338.
3 Verhelst J, Abs R: Hyperprolactinemia: pathophysiology and management. Treat Endocrinol 2003;2:23–32.
4 Casanueva FF, Molitch ME, Schlechte JA, Abs R, Bonert V, Bronstein MD, Brue T, Cappabianca P, Colao A, Fahlbusch R, Fideleff H, Hadani M, Kelly P, Kleinberg D, Laws E, Marek J, Scanlon M, Sobrinhho LG, Wass JA, Giustina A: Guidelines of the Pituitary Society for the diagnosis and management of prolactinomas. Clin Endocrinol (Oxf) 2006;65:265–273.
5 Schlechte JA: Prolactinoma. N Eng J Med 2003;349:2035–2041.
6 Schlechte JA, Dolan K, Sherman B, Luciano A: The natural history of untreated hyperprolactinemia. J Clin Endocrinol Metab 1989;2035–2041.
7 Walsh Jo adn Pullan PT: Hyperprolactinaemia in males: a heterogeneous disorder. Aust NZ J Med 1980;13:241–247.
8 Pinzone JJ, Katznelson I, Danila DC, Pauler DK, Miller CS, Klibanski A: A primary medical therapy for micro- and macroprolactinomas in men. J Clin Endocrinol Metab 2000;85:3053–3057.
9 Colao A, Di Sarno A, Cappablanca P, Briganti F, Pivonello R, Di Somma C, Faggiano A, Biondi B, Lombardi G: Gender differences in the prevalence, clinical features and response to cabergoline in hyperprolactinemia. Eur J Endocrinol 2003;148:325–331.
10 Danila DC, Klibansli A: Prolactin secreting pituitary tumors in men. Endocrinologist 2001;11:105–111.
11 Berezin M, Shimon I, Hadani M: Prolactinoma in 53 men: Clinical characteristics and models of treatment (male prolactinoma). J Endocrinol Invest 1995;18:436–441.
12 Delgrange E, Trouillas J, Maiter D, Donckier J, Tourniaire J: Sex-related difference in the growth of prolactinomas: a clinical and proliferation marker study. J Clin Endocrinol Metab 2000;85:3053–3057.
13 Trouillas J, Delgrange E, Jouanneau E, Maiter D, Guigard MP, Donckier J, Perrin G, Jan M, Tourniaire J: Prolactinoma in man: clinical and histological characteristics. Ann Endocrinol 2000;61:253–257.
14 Calle-Rrodriguez R, Giannini C, Scheithauer B, Lloyd R: Prolactinomas in male and female patients: a comparative clinicopathologic study. Mayo Clin Proc 1998;73:1046–1052.
15 Newton CJ: Estrogen receptor blockade by the pure antiestrogen, ZM 182780, induces death of pituitary tumour cells. J Steroid Biochem Mol Biol 1995;55:327–336.
16 Haug E, Gautvik KM: Effects of sex steroids on prolactin secreting rat pituitary cells in culture. Endocrinology 1976;99:1482–1489.
17 Ishida M, Takahashi W, Itoh S, Shimodaira S, Maeda S, Arita J: Estrogen actions on lactotroph proliferation are independent of a paracrine interaction with other pituitary cell types: a study using lactotroph-enriched cells. Endocrinology 2007l;148:3131–3139.
18 Hentges S, Sarkar DK: Transforming growth factor-beta regulation of estradiol-induced prolactinomas. Front Neuroendocrinol 2001;22:340–615.
19 Gillam MP, Middler S, Freed DJ, Molitch ME: The novel use of very high doses of cabergoline and a combination of testosterone and an aromatase inhibitor in the treatment of a giant prolactinoma. J Clin Endocrinol Metab 2002;87:4447–4451.
20 Cicarelli A, Guerra E, De Rosa M, Milone F, Zarrilli S, Lombardi G: PRL Secreting adenomas in male patients. Pituitary 2005;8:39–42.
21 Colao A, Vitale G, Cappabianca P, Briganti F, Ciccarelli A, De Rosa M, Zarrilli S, Lombardi G: Outcome of cabergoline treatment in men with prolactinoma: effects of a 24-month treatment on prolactin levels, tumor mass, recovery of pituitary function, and semen analysis.. J Clin Endocrinol Metab 2004;89:1704–1711.

22 Leonard MP, NickelCJ, Marales A: Hyperprolactinemia and importance: why, when and how to investigate. J Urol 1989;142:992–994.
23 Kadioglu P, Oral G, Sayitoglu M, Erensoy N, Senel B, Gaziogllu N, Sav A, Cetin G, Ozbek U: Arometase cytochrome P450 enzyme expression in human pituitary. Pituitary 2008;11:29–35.
24 Nolan LA, Levy A: The effects of testosterone and oestrogen on gonadectomised and intact male rat anterior pituitary mitotic and apoptoctic activity. J Endocrinol Invest 2006;188:387–389.
25 Carretero J, Burks DJ, Vázquez G, Rubio M, Hernández E, Bodego P, Vázquez R: Expression of aromatase P450 is increased in spontaneous prolactinomas of aged rats 3. Pituitary 2002;5:5–10.
26 Thapar K, Yamada Y, Scheithauer B, Kovacs K, Yamada S, Stefaneanu L: Assessment of mitotic activity in pituitary adenomas and carcinomas. Endocr Pathol 1996;7:215–221.
27 Gaffey TA, Scheithauer BW, Lloyd RV, Burger PC, Robbins P, Fereidooni F, Horvath E, Kovacs K, Kuroki T, Young WF Jr, Sebo TJ, Riehle DL, Belzberg AJ: Corticotroph carcinoma of the pituitary: a clinicopathological study. Report of four cases. J Neurosurg 2002;96:352–360.
28 Crusius PS, Forcelini CM, Mallmann A, Silveira DA, Lersch E, Seibert C, Crusius M, Carazzo Ch, Crusius CU, Goellner E: Metastatic prolactinoma: case report with immunohistochemical asseessment for p53 and Ki-67 antigens. Arq. Neuro-Psiquiatr. 2005;63:864–869.
29 Thapar K, Kovacs K, Scheithauer BW, Stefaneanu L, Horvath E, Pernicone PJ, Murray D, Laws ER Jr: Proliferative activity and invasiveness among pituitary adenomas and carcinomas: an analysis using the MIB-1 antibody. Neurosurgery 1996;38:99–107.
30 Wierinckx A, Auger C, Devauchelle P, Reynaud A, Chevallier P, Jan M, Perrin G, Févre-Montange M, Rey C, Figarella-Branger D, Raverot G, Melin MF, Lachuer J, Trouillas J: A diagnostic marker set for invasion, proliferation, and aggressiveness of prolactin pituitary tumor. Endoc Relat Cancer 2008;14: 987–990.
31 Verhelst J, Abs R, Maiter D, Vandeweghe M, Velkeniers B, Mockel J, Lamberigts G, Petrossians P, Coremans P, Mahler Ch, Stevenaert A, Verlooy J, Beckers A: Cabergoline in the treatment of hyperprolactinemia: a study in 455 patients. J Clin Endocrinol Metab 1999;84:2518–2522.
32 Molitch ME: Pharmacologic resistance in prolactinoma patients. Pituitary 2005;8:43–52.
33 Gürlek A, Karavitaki N, Ansorge O, Wass,JA: What are the markers of aggressiveness in prolactinomas? Changes in cell biology, extracellular matrix components, angiogenesis and genetics. Eur J Endocrinol 2007;156:143–153.
34 Corsello SM, Ubertini M, Altomare RM, Lovicu MG, Rota CA, Colosimo C: Clinical giant prolactinomas in men: efficacy of cabergoline treatment. Endocrinology 2003;58:662–670.
35 Colao A, Di Sarno A, Guerra E, De Leo M, Mentone A, Lombardi G: Drug insight: cabergoline and bromocriptine in the treatment of hyperprolactinemia in men and women. Nat Pract Endocrinol Metab 2006;2:200–204
36 Beverly MK, Molitch M, Vance L, Cannistraro K, Davis K, Simons JA, Schoenfelder JR, Klibansky A: Treatment of prolactin-secreting macroadenomas with the once-weekly dopamine agonist cabergoline. J Clin Endocrinol Metab 1996;81:2338–2343.
37 Biller BM: Medical therapy for prolactinomas and somatotroph adenomas. Endocr Pract 1996;2:333–337.
38 Colao A, Di Sarno A, Sarnacchiaro F, Ferone D, Di Renzo G, Merola B, Annunziato L, Lombardi G: Prolactinomas resistent to standard dopamine agonist respond to chronic cabergoline treatment. J Clin Endocrinol Metab 1997;82:876–883.
39 Ferrari CI, Abs R, Bevan JS, Ciccarelli E, Motta T, Mucci M, Muratori M, Musatti L, Verbessem G, Scanlon MF: Treatment of macroprolactinoma with cabergoline: a study of 85 patients. Clin Endocrinol (Oxf) 1997;46:409–413.

Patricia Fainstein Day
Endocrine and Nuclear Medicine Unit Hospital Italiano
Buenos Aires, Gascón 450
Buenos Aires, CP 1181 (Argentina)
Tel. +54 11 49590200, Fax +54 49590323, E-Mail patricia.fainstein@hospitalitaliano.org.ar

Neurotransmitter Modulation of the GHRH-GH Axis

Isabel García-Tornadu[a] · Gabriela Risso[a] ·
Maria Ines Perez-Millan[a] · Daniela Noain[b] ·
Graciela Diaz-Torga[a] · Malcolm J. Low[c] · Marcelo Rubinstein[b] ·
Damasia Becu-Villalobos[a]

[a]Instituto de Biología y Medicina Experimental, CONICET, and [b]Instituto de Investigaciones en Ingeniería Genética y Biología Molecular, CONICET and University of Buenos Aires, Buenos Aires, Argentina; [c]Center for the Study of Weight Regulation and Department of Behavioral Neuroscience, Oregon Health and Science University, Portland, Oreg., USA

Abstract

The role of dopaminergic receptors in the control of GH release remains controversial. The dopamine receptor 2 (D2R) knockout mouse represents a useful model to study the participation of the D2R on growth and GHRH-GH regulation. These knockout mice have hyperprolactinemia and lactotrope hyperplasia, but unexpectedly, they are also growth retarded. In D2R knockout mice there is a significant decrease in somatotrope population, which is paralleled by decreased GH content and output from pituitary cells. The sensitivity of GHRH-induced GH and cAMP release is similar between genotypes, even though the response amplitude is lower in knockouts. We point to an involvement of D2R signaling at the hypothalamic level as dopamine did not release GH acting at the pituitary level, and both somatostatin and GHRH mRNA expression are altered in knockout mice. The similarity of the pituitary defect in the D2R knockout mouse to that of GHRH deficient models suggests a probable mechanism. Loss of dopamine signaling via hypothalamic D2Rs at a critical age may cause inadequate GHRH secretion subsequently leading to inappropriate somatotrope lineage development. Furthermore, GH pulsatility, which depends on a regulated temporal balance between GHRH and somatostatin output might be compromised in D2R knockout mice, leading to lower IGF-I, and growth retardation.

Copyright © 2010 S. Karger AG, Basel

Dopamine is the most abundant catecholamine in the brain. Its involvement and importance as a neurotransmitter and neuromodulator in the regulation of different physiological functions in the central nervous system is well known, and deregulation of the dopaminergic system has been linked with Parkinson's disease, Tourette's syndrome, schizophrenia, attention deficit hyperactive disorder, drug addiction, obesity and generation of pituitary tumors.

Dopamine exerts its action by binding to specific membrane receptors, which belong to the family of seven transmembrane domain G-protein-coupled receptors. Five distinct dopamine receptors have been isolated, characterized and subdivided into two subfamilies, D1- and D2-like, on the basis of their biochemical and pharmacological properties. The D1-like subfamily comprises D1R and D5R, while the D2-like includes D2R, D3R and D4R. D1 receptors are coupled to stimulatory G proteins, and the D2 subtype to inhibitory Gi/Go proteins. The best-described effects mediated by dopamine acting on D2Rs are the inhibition of the cAMP pathway and modulation of Ca^{2+} signaling [1].

In brain tissues the D2R is expressed predominantly in the caudate putamen, olfactory tubercle and nucleus accumbens. It is also expressed in the substantia nigra pars compacta and in the ventral tegmental area. These are the anatomical regions that give rise to long dopaminergic fibers (A10 and A9), indicating that the D2Rs have a presynaptic location. In contrast, D1-like receptors are exclusively postsynaptic. The D2R is expressed in two isoforms (long and short), and mRNA analysis of the two isoforms has shown that D2R-L is the most abundantly expressed. Outside the brain the D2R is also localized in the retina, kidney, vascular system and pituitary gland.

At the pituitary level, dopamine acting on D2Rs inhibits prolactin secretion from lactotropes, and αMSH secretion from melanotropes. On the other hand, the role of dopaminergic receptors in the control of GH release remains controversial in several respects, i.e. the direction of action (stimulatory or inhibitory) and the species differences encountered.

Dopamine and GHRH-GH Regulation

Inhibitory as well as stimulatory effects of the amine have been reported on plasma levels of GH in vivo depending upon the experimental conditions used [2]. This may be explained by the ability of dopamine to release both GHRH and somatostatin from the rat hypothalamus [3]. In particular, it has been suggested that dopamine receptors can mediate the stimulation by dopamine of GH, provided other neural inhibitory inputs to the pituitary are removed. L-DOPA stimulates GH secretion in vivo [2], and apomorphine a central dopamine receptor agonist stimulates GH secretion. However, the GH stimulatory action of L-DOPA does not appear to be mediated via dopamine receptors as specific blockade of these receptors with antidopaminergic drugs does not alter the GH response [4, 5]. Instead, L-DOPA's effects appear to depend on conversion to noradrenaline or adrenaline, as alpha-adrenoceptor blockade with phentolamine disrupts the GH response to L-DOPA [6].

In vitro, positive as well as negative GH responses to the catecholamine have been described in pituitary cells [2, 7].

Dopamine in Acromegaly Treatment

On the other hand, dopamine agonists have been largely used for the treatment of pituitary tumors, particularly prolactinomas but also in acromegaly, and the responsiveness seems to depend on the expression of D2Rs on tumor cells [8].

Somatostatin Sst2A receptors and D2Rs are frequently co-expressed in adenomas from acromegalic patients. The additive effect of dopamine and somatostatin agonists in lowering GH suggests that the combination of somatostatin and dopamine analogues might be useful in selected patients. Chimeric molecules that are able to bind to both somatostatin and dopamine receptors are being developed for the treatment of acromegaly [9, 10]. The mechanism(s) by which such ligands may act are still unknown. One possible explanation of their increased potency could be through their ability to induce oligodimerization of the receptors at the cell membrane level, and modify, in a ligand-specific manner, the subsequent trafficking and recycling of the receptors [11].

In vitro experiments demonstrated that D2R immunoreactivity in adenomas from acromegalic patients positively correlated with the in vitro GH and PRL suppression by quinagolide in primary cultures from the pituitary adenomas [12]. However, D2R expression was not correlated with the in vivo GH response to quinagolide, suggesting that the in vivo sensitivity of acromegalic patients to dopamine might be affected by other mechanisms, for example antiangiogenesis.

Dopamine and Growth

With regard to a possible role of the dopaminergic system in growth, it has been shown that GH deficient children increase their growth velocity after 6 months of levodopa treatment, even though the possible intervention of the adrenergic system was not tested [13]. On the other hand it has been described that a group of children with idiopathic short stature, had high frequencies of the A1 allele of the D2R, indicating a polymorphism of the receptor [14].

The D2R knockout mouse represents a useful model to study the participation of the D2R on growth and GHRH-GH regulation. As pituitary D2Rs are mandatory for dopamine inhibition of prolactin synthesis and release, as well as lactotrope proliferation, knockout mice have chronic hyperprolactinemia, lactotrope hyperplasia [15, 16], and after 16 months of age highly vascularized adenomas develop, especially in females, but also in males [17]. Unexpectedly, these mice were also growth retarded evidencing an alteration in the GH-IGF-I axis [18].

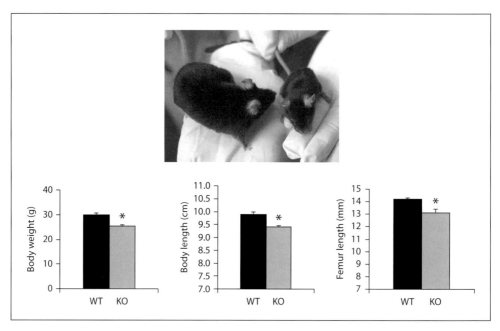

Fig. 1. Below: Body weight, body and femur length in wild-type (WT) and D2R knockout (KO) male mice at 6 months of age. Average ± SEM. Modified from Diaz-Torga et al. [18]. Upper panel representative photograph. * p < 0.05 vs. respective wild-type.

The D2R Knockout Mouse, a Dwarf Mouse

In wild-type and D2R knockout mice body weight at birth was similar, but growth retardation was evidenced starting on the second month of life. Growth retardation was especially evident in male mice, females were smaller in the first months and there was a growth catch up in the 3rd or 4th month. In males there was an overall body weight decrease of 15%.

When body growth gain was determined, it became evident that in D2R knockout male and female mice maximal growth retardation compared with that in wild-type mice, occurred during the first half of the second month of life, and thereafter animals grew normally.

Body length and the rate of skeletal maturation recapitulated the genotypic dimorphic pattern demonstrated for body weight (fig. 1). These results suggested that the D2R was involved, albeit indirectly, in body growth. A chronic treatment with recombinant GH in the first month of life reversed the body weight decrease, indicating that peripheral sensitivity to GH was maintained [unpubl. results].

Average serum GH levels in wild-type male and female mice were high during the first month of life and decreased to adult levels by 3 months of age. The distribution profile of these random GH measurements between 1 and 2 months of age revealed

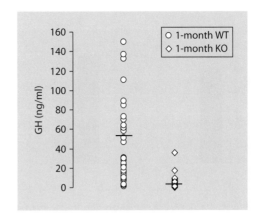

Fig. 2. Individual GH measurements in wild-type (WT) and D2R male knockout mice (KO), at 1 month of age. n = 54 and 43.

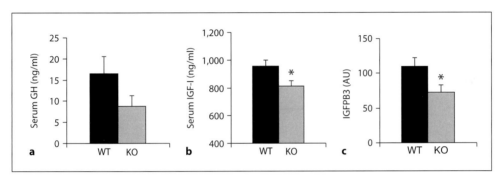

Fig. 3. Serum GH (**a**), IGF-I (**b**) and IGF-binding protein 3 (IGFBP3) (**c**). Adapted from Diaz-Torga et al. [18]. * $p < 0.05$ versus respective wild-type mice.

a high variability, probably reflecting an exaggerated pulsatility compared with adult values (fig. 2). In contrast, in female and male knockout mice GH levels were not increased during the first months of age.

This developmental period is characterized by a low effect of the somatostatin inhibitory control at the pituitary level, as well as low expression of hypothalamic somatostatin [19, 20]. The lack of increased GH levels in the first month of life had long-lasting consequences, as IGF-I levels as well as IGFBP-3 were low in adult knockout mice, even though serum GH levels were not different between genotypes in adult mice (fig. 3). This was the first evidence indicating that the D2R is involved in GH release in the first months of life [18].

The assertion was further supported by the acute GH-lowering effect of a D2R antagonist (sulpiride) in 1-month-old wild-type mice [18]. A D1R antagonist was ineffective. The effect of sulpiride was lost as the animal matured, emphasizing the importance of a low somatostatin tone to permit the unfettered effect of dopamine on GH release.

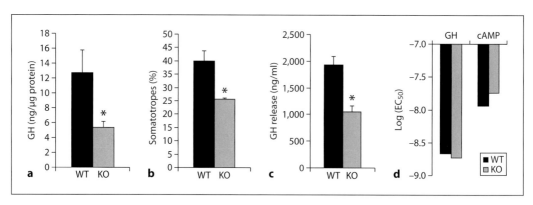

Fig. 4. Pituitary GH concentration (**a**), percentage of somatotrope cells in relation to total pituitary cells (**b**), and GH secretion from primary pituitary cultures of wild-type (WT) and D2R knockout (KO) mice (**c**). **d** Effective GHRH concentration (log M) which produces 50% increase in GH or cAMP in primary pituitary cultures of wild-type (WT) and D2R knockout (KO) mice. * $p < 0.05$ vs. respective wild-type. Adapted from Garcia-Tornadu et al. [21].

Even though GH levels in adult animals were not different between genotypes, there was a marked reduction in somatotrope number in knockouts, indicating a decreased somatotrope population size (fig. 4b) [21]. This result was paralleled by decreased pituitary GH concentration and GH secretion from pituitary cells cultured in vitro (fig. 4a, c).

In spite of the reduction in somatotrope cell number in knockouts, the functional capacity of somatotropes was not impaired, as similar dose response curve of GHRH induced GH release was observed in both genotypes (fig. 4d), even though GH net secretion was lower, in general, proportional to the low pituitary GH cell number [21]. Therefore, total GH response per pituitary was reduced, and this could account for lower IGF-I and IGFBP-3 observed in adult knockout mice [18].

This suggested that the mitotic capacity of somatotropes is very sensitive to alterations in neonatal GHRH action while the maintenance of the GH biosynthetic and secretory processes has less sensitivity to such changes.

GHRH-R protein in pituitary membranes from knockout mice was reduced to 46% of the level found in wild-type mice [21], a percentage which was higher than the reduction of somatotropes (35 %). In accordance, GHRH- induced cAMP generation was also decreased in knockouts, but the dose sensitivity was similar (fig. 4d). Somatostatin control of basal and GHRH- or ghrelin-stimulated GH release was similar between genotypes, even though D2Rs and the somatostatin receptor SSTR5 interact physically through hetero-oligomerization in neurons to create a novel receptor with enhanced functional activity [11].

In order to determine if pituitary D2Rs were involved in GH release, the effect of dopamine was tested in cultured pituitary cells. Dopamine did not modify GH acting at the pituitary level either in 1-month-old or adult mice [21]. To this regard, it has been documented that dopamine D1Rs and not D2Rs participate in GH release at the pituitary level [7], and that D2Rs are found mainly in lactotropes. Therefore, an involvement of D2R signaling at the hypothalamic level was inferred.

GH pulsatility originates at the hypothalamic level. In rodents, males exhibit narrow GH pulses with a frequency of about one pulse every 3–4 h, and prolonged nadir values below 1–2 ng/ml. Female rats exhibit relatively broader pulses with an irregular frequency and nadir values of 5–20 ng/ml. This sexually differentiated pattern is determined by the complex interplay between GHRH and somatostatin [22].

As periodical sampling is very difficult in mice, an indirect parameter of GH pulsatility has been commonly used: measurement of the major urinary proteins (MUP). MUPs (20 kDa) represent the major protein component of mouse urine. MUP expression requires pulsatile occupancy of liver GH receptors, and adult males secrete more than 3 times as much MUP as do females [23].

These proteins are synthesized in the liver, secreted through the kidneys, and excreted in urine in milligram quantities per milliliter. This abundant protein excretion is thought to play a role in chemo-signaling between animals to coordinate social behavior [24].

Knockout mice had lower MUPs, and the difference could be caused by a different pattern of GHRH or somatostatin release and action. For example, in the lit/lit mouse, which has a point mutation in the GHRH-R gene, MUPs are also decreased [25].

Further evidence of a central participation of D2Rs was the decrease in hypothalamic GHRH mRNA found in knockout compared to wild-type mice. This decrease was not simply secondary to GH deficiency or dwarfism because other transgenic lines with dwarfism and pituitary GH deficiency, caused by a primary somatotrope defect, showed the expected increase in GHRH expression [26].

Reduced levels of GHRH within the hypothalamus or GHRH action at the pituitary level during a critical developmental window have a long lasting impact on body weight [27–29], and induce an inadequate clonal expansion of the somatotrope population. The requirement of GHRH for the normal development of the somatotrope lineage is evident from studies examining the etiology of growth retardation in the spontaneous mutant mouse *lit/lit*, in which somatotropes fail to proliferate normally, resulting in a mature pituitary containing a limited number of GH cells [27]. Humans with mutations in the GHRH receptor show that defective GHRH receptor signaling results in profound, selective GH deficiency and dwarfism [30].

Furthermore, experimental ablation or inhibition of GHRH by chemical or immunological means [31, 32] provides strong circumstantial support for the notion that both acute and chronic GH release is strongly dependent on the proper functioning of arcuate GHRH neurons.

No inactivating mutations or deletions in the GHRH gene have yet been reported in human subjects, but GHRH deficiency has been described in rodents as part of more complex phenotypes resulting from deletion of other genes, or after the expression of human GH (hGH) transgenes in central nervous system to inhibit GHRH expression [33]. Recently, a report describing a targeted disruption of the GHRH gene has confirmed directly the requirement for GHRH for normal growth and GH production [34].

At birth and during the first two weeks of life there was no difference in pituitary GH concentration between genotypes. This is consistent with the GHRH-independent somatotrope development described in *lit/lit* mice [27], or the similar growth rate in mice with a partial disruption of the GHRH gene (GHRH-M2) mice during the first weeks [35], and data from animals which overexpress hGH reporter gene driven by a potent promoter, in which the effect of excess GH is only apparent after 3 weeks of age [36]. Furthermore, congenital absence of the human pituitary gland does not result in abnormal birth or newborn weight [37], indicating that fetal and early postnatal development may occur independently of GH. But the requirement of GHRH for the normal development of the somatotrope lineage after birth has been clearly demonstrated as exemplified above.

On the other hand, hypothalamic somatostatin mRNA was increased in knockout mice. As D2Rs are inhibitory, the primary effect of disruption of D2Rs may be the lack of dopamine inhibition of somatostatin neurons. In this regard, it has been described that dopamine neurons in the periventricular nucleus are close to somatostatin neurons [38]. The decrease in hypothalamic GHRH content observed in knockout mice might result from the increase in somatostatin, as it has been shown that somatostatin neurons innervate GHRH neurons, and decrease GHRH expression [39].

In line with our findings, it has been described that neonatal administration of octreotide, a long-lasting somatostatin analogue decreases growth rate, hypothalamic GHRH, and sexual differentiation of GH pulsatilily [40]. Even though the number of arcuate GHRH mRNA-containing neurons was not affected by the octreotide treatment, GHRH mRNA levels per neuron were decreased by 30%, and median eminence GHRH stores by 50%.

Is There Any Clinical Significance to Our Findings?

With regard to a possible role of the dopaminergic system in growth, it has been shown that a group of children with idiopathic short stature, had high frequencies of the A1 allele of the D2R, indicating a polymorphism of the receptor. The alterations in the dopaminergic system encountered were a low binding capacity for dopamine and reduced dopaminergic function [41]. In these children there was a mild GH deficiency, decreased nocturnal GH secretion, slightly retarded bone maturation, and low blood levels of IGF-I. Therefore D2Rs might participate in some cases of idiopathic short stature.

On the other hand, D2Rs may be involved in altered hormone secretion in chronic treatment with antidopaminergic drugs. In this regard, it has been described that during neuroleptic treatment of schizophrenic patients GH nocturnal rise is blunted, and that this effect is related to the D2R-binding capacity of neuroleptics used [42].

Conclusions

Neural networks which control GH secretion participate in the fine tuning of the GHRH-GH axis. Hypothalamic D2R modulation of GHRH or somatostatin neurons, whether directly or indirectly, may be important in the regulation of the GH axis. We point to an inhibitory effect of dopamine acting on D2Rs on somatostatin. Lack of D2Rs increases somatostatin expression, and as somatostatin is inhibitory to GHRH, this neuropeptide is decreased. If this alteration takes place early in development, altered secretion of GHRH might lead to an inappropriate somatotrope lineage development. Furthermore, GH pulsatility, which depends on an adequate temporal balance between GHRH and somatostatin output might be compromised in D2R knockout mice, finally leading to lower IGF-I, and growth retardation.

Acknowledgements

This work was supported by grants from CONICET, Fundación Alberto J. Roemmers, Fundación Fiorini, and Agencia Nacional de Promoción Científica y Técnológica, Buenos Aires, Argentina (DBV).

References

1 Missale C, Nash SR, Robinson SW, Jaber M, Caron MG: Dopamine receptors: from structure to function. Physiol Rev 1998;78:189–225.
2 Muller EE, Locatelli V, Cocchi D: Neuroendocrine control of growth hormone secretion. Physiol Rev 1999;79:511–607.
3 Kitajima N, Chihara K, Hiromi H, Okimura Y, Fujii Y, Sato M, Shakutsui S, Watanabe M, Fufita T: Effects of dopamine on immunoreactive growth hormone-releasing factor and somatostatin secretion from rat hypothalamic slices perfused in vitro. Endocrinology 1989;124:69–76.
4 Massara F, Camanni F: Effect of various adrenergic receptor stimulating and blocking agents on human growth hormone secretion. J Endocrinol 1972;54: 195–206.
5 Masala A, Delitala G, Alagna S, Devilla L: Effect of pimozide on levodopa-induced growth hormone release in man. Clin Endocrinol (Oxf) 1977;7:253–256.
6 Cammani F, Massara F: Phentolamine inhibition of human growth hormone secretion induced by L-DOPA. Horm Metab Res 1972;4:128.
7 Bluet-Pajot MT, Mounier F, Durand D, Kordon C: Involvement of dopamine D1 receptors in the control of growth hormone secretion in the rat. J Endocrinol 1990;127:191–196.
8 Ferone D, de Herder WW, Pivonello R, Kros JM, van Koetsveld PM, de Jong T, Minuto F, Colao A, Lamberts SW, Hofland LJ: Correlation of in vitro and in vivo somatotropic adenoma responsiveness to somatostatin analogs and dopamine agonists with immunohistochemical evaluation of somatostatin and dopamine receptors and electron microscopy. J Clin Endocrinol Metab 2008;93:1412–1417.
9 Saveanu A, Jaquet P, Brue T, Barlier A: Relevance of coexpression of somatostatin and dopamine D2 receptors in pituitary adenomas. Mol Cell Endocrinol 2008;286:206–213.
10 Saveanu A, Lavaque E, Gunz G, Barlier A, Kim S, Taylor JE, Culler MD, Enjalbert A, Jaquet P: Demonstration of enhanced potency of a chimeric somatostatin-dopamine molecule, BIM-23A387, in suppressing growth hormone and prolactin secretion from human pituitary somatotroph adenoma cells. J Clin Endocrinol Metab 2002;87:5545–5552.
11 Rocheville M, Lange DC, Kumar U, Patel SC, Patel RC, Patel YC: Receptors for dopamine and somatostatin: formation of hetero-oligomers with enhanced functional activity. Science 2000;288:154–157.

12 Ferone D, Pivonello R, Lastoria S, Faggiano A, Del Basso de Caro ML, Cappabianca P, Lombardi G, Colao A: In vivo and in vitro effects of octreotide, quinagolide and cabergoline in four hyperprolactinaemic acromegalics: correlation with somatostatin and dopamine D2 receptor scintigraphy. Clin Endocrinol (Oxf) 2001;54:469–477.

13 Huseman CA, Hassing JM: Evidence for dopaminergic stimulation of growth velocity in some hypopituitary children. J Clin Endocrinol Metab 1984;58: 419–425.

14 Miyake H, Nagashima K, Onigata K, Nagashima T, Takano Y, Morikawa A: Allelic variations of the D2 dopamine receptor gene in children with idiopathic short stature. J Hum Genet 1999;44:26–29.

15 Kelly MA, Rubinstein M, Asa SL, Zhang G, Saez C, Bunzow JR, Allen RG, Hnasko R, Ben-Jonathan N, Grandy DK, Low MJ: Pituitary lactotroph hyperplasia and chronic hyperprolactinemia in dopamine D2 receptor-deficient mice. Neuron 1997;19:103–113.

16 Cristina C, García-Tornadú I, Diaz-Torga G, Rubinstein M, Low MJ, Becu-Villalobos D: The dopaminergic D2 receptor knockout mouse: an animal model of prolactinoma. Front Horm Res 2006; 35:50–63.

17 Asa SL, Kelly MA, Grandy DK, Low MJ: Pituitary lactotroph adenomas develop after prolonged lactotroph hyperplasia in dopamine D2 receptor-deficient mice. Endocrinology 1999;140:5348–5355.

18 Diaz-Torga G, Feierstein C, Libertun C, Gelman D, Kelly MA, Low MJ, Rubinstein M, Becu-Villalobos D: Disruption of the D2 dopamine receptor alters GH and IGF-I secretion and causes dwarfism in male mice. Endocrinology 2002;143:1270–1279.

19 Simonian SX, Murray HE, Gillies GE, Herbison AE: Estrogen-dependent ontogeny of sex differences in somatostatin neurons of the hypothalamic periventricular nucleus. Endocrinology 1998;139:1420–1428.

20 Cuttler L, Welsh JB, Szabo M: The effect of age on somatostatin suppression of basal, growth hormone (GH)-releasing factor-stimulated and dibutyril adenosine 3′-5′-monophosphate-stimulated GH release from rat pituitary cells in monolayer culture. Endocrinology 1986;119:152–158.

21 Garcia-Tornadu I, Rubinstein M, Gaylinn BD, Hill D, Arany E, Low MJ, Diaz-Torga G, Becu-Villalobos D: GH in the dwarf dopaminergic D2 receptor knockout mouse: somatotrope population, GH release, and responsiveness to GH-releasing factors and somatostatin. J Endocrinol 2006;190:611–619.

22 Waxman DJ, O'Connor C: Growth hormone regulation of sex-dependent liver gene expression. Mol Endocrinol 2006;20:2613–2629.

23 Norstedt G, Palmiter R: Secretory rhythm of growth hormone regulates sexual differentiation of mouse liver. Cell 1984;36:805–812.

24 Logan DW, Marton TF, Stowers L: Species specificity in major urinary proteins by parallel evolution. PLoS ONE 2008;3:e3280.

25 al Shawi R, Wallace H, Harrison S, Jones C, Johnson D, Bishop JO: Sexual dimorphism and growth hormone regulation of a hybrid gene in transgenic mice. Mol Endocrinol 1992;6:181–190.

26 McGuinness L, Magoulas C, Sesay AK, Mathers K, Carmignac D, Manneville JB, Christian H, Phillips JA III, Robinson IC: Autosomal dominant growth hormone deficiency disrupts secretory vesicles in vitro and in vivo in transgenic mice. Endocrinology 2003;144:720–731.

27 Lin SC, Lin CR, Gukovsky I, Lusis AJ, Sawchenko PE, Rosenfeld MG: Molecular basis of the little mouse phenotype and implications for cell type-specific growth. Nature 1993;364:208–213.

28 Cella SG, Locatelli V, Broccia ML, Menegola E, Giavini E, De Gennaro Colonna V, Torsello A, Wehrenberg WB, Muller EE: Long-term changes of somatotrophic function induced by deprivation of growth hormone-releasing hormone during the fetal life of the rat. J Endocrinol 1994;140:111–117.

29 Robinson GM, Spencer GS, Berry CJ, Dobbie PM, Hodgkinson SC, Bass JJ: Evidence of a role for growth hormone, but not for insulin-like growth factors I and II in the growth of the neonatal rat. Biol Neonate 1993;64:158–165.

30 Maheshwari HG, Silverman BL, Dupuis J, Baumann G: Phenotype and genetic analysis of a syndrome caused by an inactivating mutation in the growth hormone-releasing hormone receptor: Dwarfism of Sindh. J Clin Endocrinol Metab 1998;83:4065–4074.

31 Wehrenberg WB, Brazeau P, Luben R, Ling N, Guillemin R: A noninvasive functional lesion of the hypothalamo-pituitary axis for the study of growth hormone-releasing factor. Neuroendocrinology 1983;36:489–491.

32 Bloch B, Ling N, Benoit R, Wehrenberg WB, Guillemin R: Specific depletion of immunoreactive growth hormone-releasing factor by monosodium glutamate in rat median eminence. Nature 1984; 307:272–273.

33 Flavell DM, Wells T, Wells SE, Carmignac DF, Thomas GB, Robinson IC: Dominant dwarfism in transgenic rats by targeting human growth hormone (GH) expression to hypothalamic GH-releasing factor neurons. EMBO J 1996;15:3871–3879.

34 Alba M, Salvatori R: A mouse with targeted ablation of the growth hormone-releasing hormone gene: a new model of isolated growth hormone deficiency. Endocrinology 2004;145:4134–4143.

35 Le Tissier PR, Carmignac DF, Lilley S, Sesay AK, Phelps CJ, Houston P, Mathers K, Magoulas C, Ogden D, Robinson IC: Hypothalamic growth hormone-releasing hormone (GHRH) deficiency: targeted ablation of GHRH neurons in mice using a viral ion channel transgene. Mol Endocrinol 2005; 19:1251–1262.
36 Mathews LS, Hammer RE, Brinster RL, Palmiter RD: Expression of insulin-like growth factor I in transgenic mice with elevated levels of growth hormone is correlated with growth. Endocrinology 1988;123:433–437.
37 Reid JD: Congenital absence of the pituitary gland. J Pediatr 1960:56:658–664.
38 Fodor M, Kordon C, Epelbaum J: Anatomy of the hypophysiotropic somatostatinergic and growth hormone-releasing hormone system minireview. Neurochem Res 2006;31:137–143.
39 Bouyer K, Loudes C, Robinson IC, Epelbaum J; Faivre-Bauman A: Sexually dimorphic distribution of sst2A somatostatin receptors on growth hormone-releasing hormone neurons in mice. Endocrinology 2006;147:2670–2674.
40 Slama A, Bluet-Pajot MT, Mounier F, Videau C, Kordon C, Epelbaum J: Effects of neonatal administration of octreotide, a long-lasting somatostatin analogue, on growth hormone regulation in the adult rat. Neuroendocrinology 1996;63:173–180.
41 Ezzat S, Walpola IA, Ramyar L, Smyth HS, Asa SL: Membrane-anchored expression of transforming growth factor-α in human pituitary adenoma cells. J Clin Endocrinol Metab 1995;80:534–539.
42 Mann K, Rossbach W, Muller MJ, Muller-Siecheneder F, Pott T, Linde I, Dittmann RW, Hiemke C: Nocturnal hormone profiles in patients with schizophrenia treated with olanzapine. Psychoneuroendocrinology 2006;31:256–264.

Damasia Becu-Villalobos
Instituto de Biología y Medicina Experimental. CONICET
Vuelta de Obligado 2490
Buenos Aires 1428 (Argentina)
Tel. +54 11 47832869, Fax +54 11 47862564, E-Mail dbecu@dna.uba.ar

iASPP: A Novel Protein Involved in Pituitary Tumorigenesis?

Emilia M. Pinto[a] · Nina Rosa C. Musolino[b] · Valter A.S. Cescato[b] · Iberê C. Soares[c] · Alda Wakamatsu[c] · Evandro de Oliveira[d] · Luiz Roberto Salgado[e] · Marcello D. Bronstein[e]

[a]Laboratório de Patologia Cardiovascular, LIM22, Departamento de Patologia, Faculdade de Medicina da Universidade de São Paulo, [b]Unidade de Neuroendocrinologia, Divisão de Neurocirurgia Funcional, Instituto de Psiquiatria, HC-FMUSP, [c]Laboratório de Investigação em Patologia Hepática, LIM14, Departamento de Patologia, Faculdade de Medicina da Universidade de São Paulo, [d]Instituto de Ciências Neurológicas, and [e]Unidade de Neuroendocrinologia, Disciplina de Endocrinologia, HC-FMUSP, São Paulo, Brazil

Abstract

Pituitary tumors can be morphologically classified as microadenomas (diameter <1 cm) or macroadenomas (>1 cm), which can be enclosed, invasive and/or expansive. Functionally, they are classified as secreting tumors and clinically non-secreting or 'non-functioning' tumors. Several molecular mechanisms have been studied acting in uncontrolled cell proliferation and the acquisition of resistance to apoptosis. A potential mechanism related to apoptosis control has been found following the isolation and characterization of the ASPP proteins family. All these proteins share sequence similarities in their C-termini, which contains their signature sequences of Ankyrin repeats, SH3 domain and proline-rich region. Recent investigations reported that the expression of iASPP mRNA and protein was increased in non-transformed cells induced to undergo apoptosis and inhibition of iASPP expression in these cells by siRNA reduced apoptosis. Thus, modulation of iASPP expression seems to be an integral part of the apoptotic response. The ASPP proteins family binds to proteins that are key players in controlling apoptosis (P53 and NFκB p65 subunit). It has been speculated that the iASPP protein product induces apoptosis by blocking NFkappaB or inhibits apoptosis by blocking P53. By either mechanism, the gene could influence the survival of precancerous lesions.

Copyright © 2010 S. Karger AG, Basel

The pituitary gland responds to diverse central and peripheral signals by undergoing reversible plastic and functional changes. The resulting hyperplasia/excess hormone production or involution/hyposecretion in pituitary cells may correlate with the ability to develop pituitary tumors [1]. Pituitary tumors, almost invariably adenomas, are of frequent occurrence, accounting for 10–15% of all intracranial neoplasms and are found incidentally in up to 25% of unselected autopsies specimens [2]. Pituitary adenomas are a heterogeneous group of tumors that, in spite of pathophysiological similarities, exhibit different features depending on their functional type [3]. They are classified as microadenomas (<10 mm) or macroadenomas (>10 mm) and clinically non-secreting (or non-functioning) adenomas or as secreting when these tumors are capable of autonomously release pituitary hormones, such as growth hormone (GH), prolactin (PRL), adrenocorticotropic hormone (ACTH), thyroid-stimulating hormone (TSH), follicle-stimulating hormone (FSH) and luteinizing hormone (LH) [3–5]. The occurrence of metastases, characterizing a pituitary carcinoma, is exceedingly rare. However, tumors with aggressive behavior leading to local invasion are relatively common [4, 5]. Mechanisms underlying changes in pituitary plasticity and their relationship to tumor development may account for the diverse genetic abnormalities observed in pituitary tumors. The spectrum of cellular changes evolve from reversible hyperplasia to a committed pituitary microadenoma and ultimately to a macroadenoma [2]. Molecular events encompass increased oncogene expression or silencing of tumor suppressor genes, pituitary and hypothalamic hormonal dysregulation, and hyperexpression of growth factors. Environmental or iatrogenic mutagenic stimuli such as external radiation or pharmacological hormone administration [6] can act upon growth autonomy, unrestrained replicative potential and evasion of apoptosis [2]. These elements can be demonstrated by immunohistochemistry and/or molecular pathology but no single factor can be used for determination of biological behavior.

P53 Tumor Supressor

Pituitary tumors are composed of monoclonal cell populations with disrupted control of replication pathways. The oncogenes and tumor suppressor genes which are common in other malignancies (i.e. jun, fos, myc, and p53) are rarely involved in the development of these tumors. The P53 tumor suppressor carries out the important task of ensuring that damaged DNA is not passed on during cell division, is capable of suppressing tumor growth through its ability to induce apoptosis or cell-cycle arrest [7]. P53 is regulated at multiple levels including post-translationally by modifications such as acetylation, phosphorylation, protein degradation and protein-protein interactions. P53 is also regulated through association with ankyrin repeat proteins such as P53-binding protein 1 (53BP1) and P53-binding protein 2 (53BP2) [8]. Although P53 is one of the most frequently mutated genes found in human cancer, the mutation frequency varies among different tumor types and is rather infrequently encountered in pituitary tumors. For instance, P53 is poorly expressed in pituitary tumors and its expression pattern is not linked to a specific phenotype [9, 10]. Recently, the ASPP family of proteins was identified, which act to direct the cell away from cell cycle arrest and towards death following P53 upregulation [8].

ASPP Family

The apoptosis-stimulating proteins of the P53 (ASPP) family consists of three members, ASPP1, ASPP2 and iASPP, that bind to proteins which are key players in controlling apoptosis (P53, Bcl-2 and RelA/p65) and cell growth (APCL, PP1). ASPP stands for apoptosis-stimulating protein of P53, and the name emphasizes the ankyrin repeats, SH3 domain, and proline-rich domains that characterize the family [8]. ASPP proteins interact with P53 and specifically enhance the DNA-binding and transactivation function of P53 on the promoters of pro-apoptotic genes in vivo [8] but not cell cycle arrest. Biochemical and genetic evidence has shown that ASPP1 and ASPP2 activate, whereas iASPP inhibits, the apoptotic function of P53 [11]. Thus, the ASPP family of proteins helps to determine how cells choose to die and may therefore be a novel target for cancer therapy [11]. ASPP2 is in fact the full-length form of the previously identified 53BP2, a P53-binding protein that contains a proline-rich sequence, four ankyrin repeats, and an SH3 domain. The crystal structure of a complex of parts of P53 and 53BP2 has revealed that the 53BP2-binding site on the central domain of P53 consists of evolutionarily conserved regions [12]. Subsequently, 53BP2 was also found to interact with other proteins, including Bcl-2 [13] and RelA/p65, a component of NFkappaB [14]. ASPP1 is slightly shorter than ASPP2 and the greatest homology between the two proteins exists at the N- and C-termini. All the residues in the SH3 and ankyrin repeat regions of ASPP2 that are required for contact with P53 are conserved in ASPP1 and, akin to, ASPP2, ASPP1 is able to bind to the DNA-binding domain of P53 [8]. The inhibitory member of the ASPP family, the iASPP protein, was identified previously as a protein of 351 amino acids that interacts with and inhibits the p65 subunit (RelA), and was thus named RelA-associated inhibitor (RAI). The iASPP protein also contains 4 consecutive ankyrin repeats and a C-terminal SH3 domain while it has been speculated that the protein product induces apoptosis by blocking NFκB or inhibits apoptosis by blocking P53. By either mechanism, the gene could influence the survival of precancerous lesions [15]. In a previous study, the expression of iASPP mRNA was increased in non-transformed lymphocytes and fibroblasts induced to undergo apoptosis by various means, such as treatment with etoposide, calcium ions, or interleukin-2 and/or serum deprivation. Treatment with etoposide also increased the content of iASPP protein and caused it to translocate to the nucleus. Inhibition of iASPP expression in lymphocytes and fibroblasts with siRNA reduced apoptosis, but treatment with the NFκB-inhibiting substance sulfasalazine relieved this dependence, suggesting a NFκB-mediated mechanism [15]. The direct interaction between iASPP/RAI and p65 (RelA) most likely contributes to the inhibition of NFκB activity through interfering with its NFκB DNA-binding activity. Thus, it may act as a novel NFκB-binding protein, causing a repression of transcriptional activity, and preventing pre-cancerous cells to survive [15].

NFkappaB

The NFκB family of transcription factors is known to be implicated in the cellular response to stress and have been characterized as participating in both pro- and anti-apoptotic pathways [16]. The activity of NFκB is tightly regulated by its interaction with inhibitory IkB proteins. In most resting cells, NFκB is sequestered in the cytoplasm by a family of inhibitors, called IkBs, such as IkBα, IkBβ, IkBε, p105, and p100 which are proteins that contain multiple copies of a sequence of ankyrin repeats. By virtue of their ankyrin repeat domains, the IkB proteins mask the nuclear localization signals of NFκB proteins and keep them sequestered in an inactive state in the cytoplasm. This interaction blocks the ability of NFκB to bind to DNA and results in the NFκB complex being primarily localized to the cytoplasm due to a strong nuclear export signal in IkB. Following exposure to inflammatory cytokines, UV light, reactive oxygen species, or bacterial and viral toxins, the NFκB signaling cascade is activated, leading to the complete degradation of IkB. This allows the translocation of unmasked NFκB to the nucleus where it binds to the enhancer or promoter regions of target genes and regulates their transcription. The iASPP protein predominantly locates in the cytoplasm, thus it may only inhibit and sequester proteins in this location, interacting with other proteins via ankyrin repeats and the SH3 domain [17].

iASPP in Pituitary Tumors

In the present study, we assessed iASPP RNAm expression by real-time PCR using the TaqMan assay which contains primers and probe specific to human iASPP mRNA. We obtained RNA samples from pituitary tumors of 21 Brazilian patients (15 females and 6 males, aged between 20 and 59 years, mean ± SD: 38 ± 13). Twenty patients underwent surgery by the trans-sphenoidal approach and one by the transcranial route. The diagnoses based on clinical, hormonal and immunohistochemical evaluations were: nonfunctioning pituitary adenoma in 7 cases, acromegaly in 5 cases, prolactinoma in 4 cases, Cushing's disease in 3 cases, FSH-secreting adenoma in 1 case and TSH-secreting adenoma in another patient. Tumor size was evaluated by magnetic resonance imaging in all patients. Macroadenoma (MAC) was disclosed in 19 subjects; microadenoma (MIC) in one while no tumor was visible in one patient with Cushing's disease (table 1). Lower levels of iASPP mRNA were detected in all subtypes of tumor, independently of morphological and functional features, compared with normal tissue. The level of expression ranged from 0.01 to 0.31 in nonfunctioning tumors (7 cases); 0.02 to 0.617 in GH-secreting tumors (5 cases); 0.04 to 0.36 in PRL-secreting tumors (4 cases); 0.02 to 0.16 in ACTH-secreting tumors (3 cases). The level observed in FSH-secreting tumors (1 case) was 0.02, and 0.08 for TSH-secreting tumors (1 case). The analysis of the expression in all cases ranged from 0.012 to 0.617 (median = 0.081) compared to normal (= 1.0) ($p < 0.01$) (fig. 1). The statistical analysis showed no relationship among the

Table 1. Patient data including immunohistochemical findings; tumor maximum diameter and the expression level of mRNA iASPP

Patient No.	Gender	Age years	Diagnosis	Findings at immunohistochemistry	Tumor type	Maximum diameter, cm	mRNA iASPP
1	M	22	acromegaly	GH, LH (focal)	MAC	2.9	0.62
2	F	33	acromegaly	GH	MAC	3	0.02
3	F	37	acromegaly	GH, PRL	MAC	2.9	0.17
4	F	37	acromegaly	GH, PRL (focal)	MAC	n.a.	0.06
5	F	29	acromegaly	GH, PRL (focal)	MAC	4.6	0.03
6	F	51	Cushing	ACTH	MAC	3.5	0.02
7	M	23	Cushing	ACTH	MIC	n.v.	0.08
8	F	24	Cushing	ACTH	MIC	0.9	0.16
9	F	37	PRL	PRL	MAC	1.4	0.08
10	F	24	PRL	PRL	MAC	2.3	0.16
11	F	31	PRL	PRL	MAC	1.1	0.36
12	M	53	PRL	PRL	MAC	4	0.04
13	F	51	NF	TSH (focal), FSH (focal), LH (focal)	MAC	3.6	0.31
14	M	58	NF	TSH (focal), FSH (focal), LH (focal)	MAC	2.8	0.17
15	M	59	NF	TSH (focal), FSH (focal), LH (focal)	MAC	3	0.02
16	F	59	NF	TSH (focal)	MAC	5.5	0.02
17	F	43	NF	TSH (focal), FSH (focal), LH (focal)	MAC	2.4	0.02
18	M	26	NF	null cell	MAC	5.2	0.01
19	F	36	NF	null cell	MAC	2.8	0.12
20	F	48	TSHoma[1]	TSH, GH (focal)	MAC	1.8	0.08
21	F	20	FSHoma[2]	TSH (focal), FSH (focal)	MAC	6	0.02

n.a. = Not accessible; n.v. = not visualized.
[1] Clinical and laboratory hyperthyroidism with normal TSH.
[2] Serum FSH and α subunit elevated with enlarged ovaries.

variables of interest (age, gender, diagnosis, maximum diameter, immunohistochemical findings, tumor type and expression levels of iASPP). We also performed a tissue array using formalin fixed paraffin embedded tissues (FFPE) from 5 normal pituitary tissue and the same tumors tissue used in a real-time RT-PCR. The LX142–3 mouse

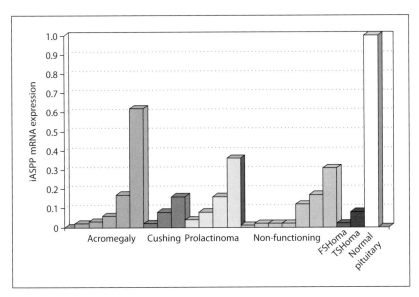

Fig. 1. iASPP mRNA expression evaluated in pituitary tumors subtypes. The white bar represents a pool of normal pituitary tissue. Lower levels of iASPP mRNA were detected in all subtypes of tumor, independently of morphological and functional features compared with normal tissue.

monoclonal antibody, reacting with iASPP protein was a generous gift from Dr. Xin Lu, London, UK. Immunostaining for this iASPP antibody revealed granular positivity in cytoplasm of the normal pituitary tissue. In opposition, none of 21 tumors displayed any iASPP staining in cytoplasm or nucleus. This data support the results observed in real time PCR. The results allow us to demonstrate that both iASPP mRNA and protein accumulation are lost in pituitary adenomas and that loss of expression of mRNA and protein of iASPP observed in all pituitary tumors subtypes compared to normal tissue suggests its participation in tumorigenesis process. The low expression of iASPP evident in all tumor subtypes compared to normal tissue also suggests that the loss or inactivation of iASPP can be an early event in the process of pituitary tumorigenesis. The iASPP protein can act on P53 and NFκB in the regulation of apoptosis. Not only are P53 and NFκB transcription factors coregulated under a variety of physiological conditions, but similarity has also been noted between crystal structures of P53 and NFκB [18]. Both proteins contain a similarly sited zinc atom that coordinates site-specific DNA binding and have similar secondary and tertiary organization. Furthermore, the p65 subunit of NFκB has been shown to bind the ankyrin protein 53BP2 [18]. Interactions between the two proteins have been noted previously and the two proteins play central roles in the control of proliferation and apoptosis. Overexpression of iASPP in mRNA level exists in acute leukemia and breast cancer with wild-type P53 [19, 20]. However, research indicates that whether ASPP family specifically regulates the apoptotic function of P53 may depend on cell background [17]. The ankyrin repeat domain appears to be a key point of interaction between diverse proteins in the NFκB and P53 superfamilies [18]. Both P53 and IkB are members of multi-gene families, and both protein families regulate apoptosis. A cooperative relationship between P53-dependent apoptosis and NFκB activation has been reported [21]. In this way iASPP might regulate apoptosis in pituitary tissue via alternative mechanisms. However, further research is needed in order to unveil the role and mechanisms of this novel protein in pituitary tumorigenesis.

References

1 Donangelo I, Melmed S: Pathophysiology of pituitary adenomas. J Endocrinol Invest 2005; 28(11 suppl int):100–105.
2 Melmed S: Mechanisms for pituitary tumorigenesis: the plastic pituitary. J Clin Invest 2003;112:1603–1618.
3 Kovacs K, Horvath E, Vidal S: Classification of pituitary adenomas. J Neurooncol 2001;54:121–127.
4 Bronstein MD, Melmed S: Pituitary tumorigenesis. Arq Bras Endocrinol Metabol 2005;49:615–625.
5 Pinto EM, Bronstein MD: Molecular aspects of pituitary tumorigenesis. Arq Bras Endocrinol Metabol 2008;52:599–610.
6 Heaney AP: Pituitary tumour pathogenesis. Br Med Bull 2005;75–76:81–97.
7 Soussi T: P53 alterations in human cancer: more questions than answers. Oncogene 2007;26:2145–2156.
8 Samuels-Lev Y, et al: ASPP proteins specifically stimulate the apoptotic function of P53. Mol Cell 2001;8:781–794.
9 Hentschel SJ, et al: P53 and MIB-1 immunohistochemistry as predictors of the clinical behavior of nonfunctioning pituitary adenomas. Can J Neurol Sci 2003;30:215–219.
10 Oliveira MC, et al: Expression of P53 protein in pituitary adenomas. Braz J Med Biol Res 2002;35:561–565.
11 Sullivan A, Lu X: ASPP: a new family of oncogenes and tumour suppressor genes. Br J Cancer 2007;96:196–200.
12 Gorina S, Pavletich NP: Structure of the P53 tumor suppressor bound to the ankyrin and SH3 domains of 53BP2. Science 1996;274:1001–1005.
13 Naumovski L, Cleary ML: The P53-binding protein 53BP2 also interacts with Bcl2 and impedes cell cycle progression at G2/M. Mol Cell Biol 1996;16:3884–3892.
14 Yang JP, et al: Identification of a novel inhibitor of nuclear factor-kappaB, RelA-associated inhibitor. J Biol Chem 1999;274:15662–15670.
15 Laska MJ, et al: Expression of the RAI gene is conducive to apoptosis: studies of induction and interference. Exp Cell Res 2007;313:2611–2621.
16 Slee EA, Lu X: The ASPP family: deciding between life and death after DNA damage. Toxicol Lett 2003;139:81–87.
17 Zhang X, et al: Identification of a novel isoform of iASPP and its interaction with P53. J Mol Biol 2007;368:1162–1171.
18 Dreyfus DH, et al: Modulation of P53 activity by IkappaBalpha: evidence suggesting a common phylogeny between kappa B and P53 transcription factors. BMC Immunol 2005;6:12.
19 Bergamaschi D, et al: iASPP oncoprotein is a key inhibitor of P53 conserved from worm to human. Nat Genet 2003;33:162–167.
20 Zhang X, et al: The expression of iASPP in acute leukemias. Leuk Res 2005;29:179–183.
21 Ryan KM, et al: Role of NF-kappa B in P53-mediated programmed cell death. Nature 2000;404:892–897.

Emilia Modolo Pinto, PhD
Laboratório de Patologia Cardiovascular, LIM22, Departamento de Patologia, Faculdade de Medicina da Universidade de São Paulo – FMUSP
Av. Dr. Arnaldo, 455, 1 andar sala 1144
01246–903 São Paulo (Brasil)
Tel. +55 11 3061-7416, Fax +55 11 3062-8098, E-Mail emiliap@plugnet.com.br

Familial Isolated Pituitary Adenoma: Evidence for Genetic Heterogeneity

Rodrigo A. Toledo · Delmar M. Lourenço Jr. · Sergio P.A. Toledo

Endocrine Genetics Unit (LIM-25), Endocrinology, Hospital das Clínicas, University of São Paulo, School of Medicine, São Paulo, Brazil

Abstract

The identification of mutations in the *Aryl hydrocarbon receptor interacting protein* (*AIP*) gene in a subset of familial isolated pituitary adenoma (FIPA) cases has recently expanded our understanding of the pathophysiology of inherited pituitary adenoma disorders. However, a genetic cause of has not yet been determined in the majority (85%) of FIPA families and half of the families with isolated familial somatotropinoma. Several studies and reviews have assessed the genetic and clinical features of *AIP*-mutated FIPA patients, which range from a complete lack of symptoms in adult/elderly individuals to large, aggressive early-onset pituitary tumors. In this study, we aimed to briefly revise the data available for the 11q13 locus and other additional loci that have been implicated in genetic susceptibility to FIPA: 2p16–12; 3q28; 4q32.3–4q33; chr 5, 8q12.1, chr 14, 19q13.4 and 21q22.1. These candidate regions may contain unidentified gene(s) that can be potentially disrupted in *AIP*-negative FIPA families. A better knowledge of these susceptibility loci may disclose modifier genes that are likely to play exacerbating or protective roles in the phenotypic diversity of *AIP*-mutated families.

Copyright © 2010 S. Karger AG, Basel

The prevalence of pituitary adenomas has been largely underestimated. Recent data indicate that about 10% of the healthy population may present with pituitary abnormalities compatible with silent adenomas [1]. Autopsy studies revealed that 6–22% of the investigated individuals had pituitary tumors [2]. A recent epidemiological study indicated that the overall rate of symptomatic pituitary adenomas in the general population is about 1 in 1,000 [3].

Although the majority of the pituitary adenomas are sporadic, up to 5% of patients may present with a familial form of the disease, which are usually associated with multiglandular neoplastic syndromes such as multiple endocrine neoplasia type 1 (MEN1) and Carney complex (CNC), or with familial isolated pituitary adenoma (FIPA), an entity that seems to involve the pituitary gland exclusively [4–6]. FIPA is defined as the occurrence of pituitary adenomas of any tumor type in ≥2 related

Table 1. Genetics of familial pituitary adenoma syndromes

Syndrome	Susceptibility genes/loci	Evidence	Pituitary disease	References
MEN1	MEN1 (11q13)	germline mutation, LOH	mostly prolactinomas, nonsecreting adenomas, GH-secreting adenomas	4, 8, 31
CNC	PPKR1A (17q22–24)	germline mutation, LOH	GH-secreting adenomas	9
	2p16 – CNC-2 oncogene (?)	positive linkage, absence of LOH, 2p16 amplification	GH-secreting adenoma	37
MEN4	CDKN1B/p27Kip	germline mutation	GH-secreting (familial) and ACTH-secreting (sporadic) adenomas**	33, 34
FIPA	AIP	germline mutation, LOH	GH-, PRL-, GH/PRL-, ACTH-, TSH-secreting and nonsecreting adenomas	10, 11, 12
	additional TSG gene at 11q13	11q13-LOH in somatotropinomas from IFS patients without AIP mutation	GH-secreting adenomas	27
	2p16 – CNC-2 oncogene (?)	suggestive/positive linkage, absence of LOH	GH-secreting adenomas	16, 39 Toledo RA, [unpubl]
	3q28	suggestive linkage lod = 1.1	GH-, PRL-, GH/PRL-secreting adenomas	40
	4q32.3–4q33	suggestive linkage lod = 3.15*		10
	Chr 5	suggestive linkage lod = 3.21*		10
	Chr 8	suggestive linkage lod = 3.08*		10
	8q12.1	suggestive linkage lod = 1.8		40
	Chr 14	suggestive linkage lod = 3.59*		10
	19q13.4	suggestive linkage lod = 2.2		40
	21q22.1	suggestive linkage lod = 1.4		40

Table 1. Continued

Syndrome	Susceptibility genes/loci	Evidence	Pituitary disease	References
MEN1-related states	CDKN2B/ p15INK4b	germline mutation (rare)	N41D and L64R (identified pathological mutations) were not associated with pituitary tumors**	35
	CDKN2C/ p18-INK4C	germline mutation (rare)	V31L (identified pathological mutation) was not associated with pituitary tumors**	
	CDKN1A/ p21CIP1	germline mutation (rare)	PRL-secreting adenoma**	
	CDKN1B/ p27Kip1	germline mutation (rare)	ATG-7G>C; P95S and stop>Q (TAA/CAA) (identified pathological mutations) were not associated with pituitary tumors**	

* After linkage disequilibrium removal, these lod scores were no longer significant [ref 10, online supplemental data].
** A complete phenotypic characterization of patients harboring mutations in *cyclin-dependent kinase inhibitor* (*CDKI*) genes is still in progress; few patients harboring these mutations have been reported so far [33–35].
Syndromes (OMIM): MEN1 = multiple endocrine neoplasia type 1 (131100); CNC = Carney complex (type 1, 160980; type 2, 605244); MEN4 = multiple endocrine neoplasia type 4 (MEN1-like) (610755); FIPA = familial isolated pituitary adenoma (102200).

family members in the absence of MEN1 or CNC [3]. Importantly, nearly all types of anterior pituitary adenomas have been described in FIPA families. Families which may be classified as 'heterogeneous FIPA' (relatives have different types of pituitary tumors) or 'homogenous FIPA' (all relatives have the same type of pituitary tumor) [6], the latter including isolated familial somatotropinoma (IFS) [7].

During the last decade, our understanding of the genetic susceptibility to familial pituitary disorders has greatly increased due to the identification of the genes responsible for MEN1 and CNC syndromes [8, 9]. Approximately 80% of patients with MEN1 and up to 60% of patients with CNC harbor heterozygous germline mutations in the *MEN1* and *PRKAR1A* genes, respectively. The loss of the remaining wild-type allele in the tumors (LOH) of MEN1 and CNC patients indicates that these genes both act as tumor suppressors [4, 5].

The first gene associated with FIPA, the *aryl hydrocarbon receptor interacting protein* (*AIP*), was recently identified [10]. Germline mutations in the *AIP* gene occur in about 15% of FIPA families and 50% of IFS families, indicating the probable existence of additional undetermined FIPA susceptibility gene(s) or genes [11]. Within the

AIP-mutated FIPA families, the penetrance of pituitary disease varies greatly, from 30% to 100%, however it is usually reduced [10–13]. In addition, a marked phenotypic diversity may also be observed among patients harboring a germline mutation in the *AIP* gene. These findings strongly suggest the presence of modifier genes that may play a role in FIPA [10–13].

In this study, we review a number of candidate loci that are likely to be potentially involved in the genetic susceptibility to FIPA. Further investigation of these loci, as well as other loci so far not identified, may greatly increase our knowledge of familial pituitary disorders.

The 11q13 Locus and the *AIP* Gene

In accordance with the Knudson's two-hit hypothesis for tumorigenesis, the identification of 11q13-LOH in somatotropinomas of IFS patients has implicated the 11q13 locus in FIPA [14, 15]. Later, Gadelha et al. [16] have documented positive linkage of 11q13 and IFS. Subsequent studies of this group were able to narrow down the candidate region to 11q13.1–11q13.3 [17, 18]. Also, the *MEN1* gene (11q13) has been screened in several FIPA families, as up to 40% of MEN1 patients may develop pituitary tumors, however no *MEN1* mutations were identified [17–21].

In 2006, Vierimaa et al. [10] combined a comprehensive single nucleotide polymorphism and expression analysis to investigate a large Finnish pedigree with pituitary adenoma predisposition, and identified the first *AIP* germline mutations in FIPA cases. The *AIP* gene, also known as *ARA9* or *XAP2*, is located at 11q13.3 and encodes the protein (AIP) that contains one FKBP-homology and three tetratricopeptide domains [22–23]. Through these domains, AIP interacts with the transcriptional factor AHR (also known as the dioxin receptor). AIP have several other partner proteins, including HSP90, PRAR, and surviving (revised in [24]).

Although the pathophysiological mechanisms of AIP are yet to be thoroughly defined, there are strong evidences to indicating that *AIP* acts as a tumor suppressor gene through cAMP regulation by interacting with type 2A and type 4A phosphodiesterases (PDE2A and PDE4A) [25, 26]. Recently, Leontiou et al. [27] have investigated several inactivating *AIP* mutations identified in FIPA patients and showed that these mutants fail to interact with PDE4A *in vitro*.

To date, more than thirty different *AIP* mutations have been identified: they are spread thoughout the gene, present incomplete penetrance and usually lead to truncated AIP protein. Although the most frequently found pituitary adenomas in *AIP* mutation carriers were GH-, GH/PRL- and PRL-secreting; ACTH-secreting and nonsecreting pituitary adenomas have also been reported [10–12]. Importantly, large and aggressive early-onset pituitary tumors have been observed in the *AIP*-mutated FIPA families. Recently, clinical, molecular and genetic features of large FIPA cohorts have been revised by Daly et al. [28, 29]. Furthermore, Cazabat et al. [30] also recently revisited the clinical and genetic aspects of FIPA and observed a genotype:phenotype correlations in the analyzed cohort: patients with *AIP* mutations that result in a truncated protein seem to develop tumors at younger ages than patients bearing a mutation which preserves the structure of the C-terminal end of the protein.

Thus, the identification of *AIP* mutations has largely broadened the spectrum of known molecular defects that cause susceptibility to familial pituitary adenoma. However, the genetic etiology of the majority of FIPA genealogies still unknown. Taken together, these data suggest that other FIPA loci/genes may exist, as discussed below.

Additional Susceptibility Genes/loci for FIPA

11q13 (Non-AIP)

It has been demonstrated that both the *MEN1* and *AIP* 11q13 genes play an important role in pituitary cell biology as *MEN1*-and *AIP*-mutated patients are prone to develop almost all types of anterior pituitary adenomas [10, 31]. The analysis of pituitary tumors from the IFS patients argued for the presence of an additional FIPA gene at 11q13, as 50% of the IFS patients with somatotropinomas presenting 11q13-LOH did not harbor an *AIP* germline mutation [27]. If this additional candidate gene does exist, 11q13 could be considered a gene *cluster* implicated in the pituitary tumorigenesis. Thus, it would be interesting to determine whether the amplitude of somatic losses at specific candidate areas of this *cluster* are associated with tumor aggressiveness.

CDKN1B/p27Kip1

Pellegata and colleagues [32, 33] have recently shown that germline mutations in the *cdkn1b* gene, which encodes the p27Kip1/cyclin-dependent kinase inhibitor 1B, are associated with the MENX syndrome in rats, a condition characterized by the occurrence of multiple MEN1- and MEN2-related tumors. In that study, the first *CDKN1B* germline mutation identified in humans was also reported. The W76X *CDKN1B* mutation was found in a family with a MEN1-like phenotype (MEN4 syndrome), involving cases with acromegaly, primary hyperparathyroidism, renal angiomyolipoma and testicular cancer [33].

Subsequently, another *CDKN1B* germline mutation was described in an apparently 'sporadic' MEN1 patient with ACTH-secreting pituitary adenoma (Cushing's disease) [34]. Recently, three other novel *CDKN1B* mutations were identified in a large cohort of patients with MEN1-related states in whom no detectable *MEN1* mutation was found [35]. Notably, none of these *CDKN1B*-mutation positive patients had developed pituitary adenomas. These data suggested that the tumoral susceptibility caused by *CDKN1B* mutations needs to be further characterized, and that the association of *CDKN1B* mutations with pituitary disease may be less prevalent than initially thought [35]. Recently, the *CDKN1B* gene was screened in an international cohort of 86 FIPA families and no significant association was observed, suggesting that *CDKN1B* may not play an important role in the etiology of FIPA [36].

Using automated direct sequencing, our laboratory has recently analyzed the *CDKN1B* gene in three additional *AIP*-negative FIPA families and no detectable mutation was found [Toledo RA, unpubl. data]. In addition to mutations in *CDKN1B*, Agarwal et al. [35] have shown that patients with MEN1-states may (rarely) harbor germline mutations in other genes encoding cyclin-dependent kinase inhibitors (CDKIs), such as *CDKN2B/p15INK4b*, *CDKN2C/p18-INK4C* and *CDKN1A/p21CIP1* [35]. Mutation in the *CDKN1A/p21CIP1* gene was associated with pituitary tumors (macroprolactinomas) in two relatives. Concluding, our knowledge about the role of the *CDKI* genes in the susceptibility to familial pituitary adenomas is still limited due to the small number of mutated patients reported so far [33–35].

2p16

Stratakis et al. [37] have identified the 2p16 region as the second locus involved in Carney complex (CNC-2), a familial tumoral endocrine syndrome that involves acromegaly in up to 10% of the cases. In further studies, this research group has assessed tumors from patients with CNC, including a GH-secreting pituitary adenoma, and documented 2p16 amplification, indicating that this region is likely to contain proto-oncogene(s) related to endocrine tumorigenesis and CNC-2 [38].

Further, Gadelha et al. [16] have shown that the 2p16/CNC-2 region is also potentially implicated in IFS, as a suggestive linkage (lod score = 2.5) was obtained for this region. Of note, no 2p16-LOH was verified in a pituitary adenoma sample from an IFS patient [16], strengthening the hypothesis that a proto-oncogene rather than a tumor suppressor may be located in this locus. Stratakis and Kirschner [39] have used the genetic data reported by Gadelha et al. [16] to recalculate the LOD scores applying different assumptions and parameters. This analysis yielded a two-point LOD score of up to 3.0 (positive) in an IFS family [39]. To the best of our knowledge, no other report on the potential role of 2p16/CNC-2 in FIPA has been published to date.

3q28, 4q32.3–4q33, Chr 5, 8q12.1, Chr 14, 19q13.4 and 21q22.1

Vierimaa et al. [10], reported that, besides 11q13.3, suggestive linkage has been observed for 4q32.3–4q33 and chromosomes 5, 8 and 14, indicating potential additional susceptibility loci [online supplemental data in ref. 10]. Although only the

11q13 region retained lod score over 3.0 after correction for linkage disequilibrium, it would be worthwhile to further analyze these regions in additional FIPA families. Further, Khoo et al. [40] have also used a genomewide strategy to study a large family with acromegaly and found suggestive linkage at 8q12.1 region previously reported by Vierimaa, as well as in three novel potential FIPA susceptibility loci: 3q28, 19q13.4, and 21q22.1. No further report has been published on these novel candidate regions.

Modifier Genes in *AIP*-Mutated Families

The knowledge that *AIP*-mutation-positive carriers may develop aggressive pituitary tumors has an important impact on the clinical management of FIPA families. Meanwhile, several studies have demonstrated that *AIP*-mutated patients, even within the same family, may present wide phenotypic variations and disease outcomes [10–13]. For example, we have described the Y268X *AIP* germline mutation in two siblings presenting with early-onset acromegaly (~20 yrs old) due to large pituitary tumors, which we have followed up in our hospital for more than 20 years. Worthwhile the *AIP* mutation was also present in a brother who were clinically asymptomatic at 41 yrs of age [12]. Furthermore, in a large FIPA family comprising 10 *AIP*-mutated relatives described by Naves et al. [13], only 3 individuals (30%) developed pituitary tumors. These data are in accordance with the original report of Vierimaa et al. [10], indicating that *AIP* mutations are associated with incomplete penetrance of pituitary disease. From a genetic point of view, this argues for the existence of modulating factors in *AIP*-mutated FIPA families.

In a previous study we have conducted, we approached possible modifier genes in FIPA [12]. To address the wide range of phenotypic variation observed in the 3 IFS relatives harboring the Y268X *AIP* mutation, we further sequenced the genes encoding the somatostatin receptors 2 and 5 *(SSTR2* and *SSTR5)*. These genes are involved in somatostatin-mediated GH suppression, and *SSTR5* variants have been associated with higher GH and IGF levels in patients with sporadic acromegaly [41], however no *SSTR2/SSTR5* variants co-segregated with acromegaly, arguing against a role in FIPA [12]. In an ongoing study of our laboratory, we had the opportunity to expand our analysis of other putative susceptibility genes in FIPA families bearing *AIP* mutations [Toledo RA, unpubl. data; partially presented at the Pituitary Today-II Meeting, 2008; Angra dos Reis, Brazil].

Recently, Khoo et al. [40] have identified the 3q28, 8q12.1, 19q13.4 and 21q22.1 regions as potential modifier loci (lod scores between 1.1 and 2.2) in a large *AIP*-mutation positive FIPA family from Borneo.

Further comprehensive investigations on the the roles of known and novel susceptibility loci are warranted to improve our knowledge of the genetic mechanisms underlying *AIP*-positive and *AIP*-negative FIPA families.

Conclusions

AIP germline mutations occur in approximately 15% of FIPA families and 50% of IFS families.

AIP germline mutations are accompanied by loss of the wild-type allele in pituitary tumors, suggesting a tumor suppressor role for *AIP*.

Patients positive for *AIP* mutations are prone to larger and early-onset pituitary tumors comparing with sporadic pituitary tumors.

AIP mutations are associated with susceptibility to nearly all types of anterior pituitary adenomas, although GH-, GH/PRL-, and PRL-secreting are observed more frequently.

The wide inter- and intrafamilial phenotypic variation and the variable (usually low) penetrance of pituitary disease in *AIP*-mutated patients suggests the existence of modifier genes that may play a role in FIPA.

There is evidence for a second FIPA gene at 11q13 (in addition to *AIP*) as only a subset of IFS patients with somatotropinomas presenting 11q13-LOH harbor a germline *AIP* mutation.

The *CDKN1B/p27Kip1* gene does not seem to play an important role in the etiology of FIPA.

Several loci with suggestive linkage to FIPA have been identified: 3q28, 4q32.3–4q33, Chr 5, 8q12.1, Chr 14, 19q13.4 and 21q22.1. These regions may contain unidentified genes that are causative of FIPA and/or genes that modulate the FIPA phenotype.

Suggestive linkage between IFS and 2p16 (CNC-2 locus) has been reported. As no 2p16-LOH was observed in the pituitary tumors of IFS and CNC patients, and amplification of 2p16 has been demonstrated in CNC tumors (including pituitary adenomas), it is likely that a proto-oncogene involved in endocrine tumorigenesis is located within this locus.

Acknowledgements

This study was supported by São Paulo State Research Foundation (FAPESP) Grant No. 08/58552–0 and the School of Medicine Foundation (FFM). R.A.T. and D.M.L. Jr. hold a FAPESP post-doc fellowship (No. 2008/58552–0 and 2009/15386-6), and SPAT is partially supported by Conselho Nacional de Desenvolvimento Científico e Tecnológico (CNPq).

References

1 Hall WA, Luciano MG, Doppman JL, Patronas NJ, Oldfield EH: Pituitary magnetic resonance imaging in normal human volunteers: occult adenomas in the general population. Ann Intern Med 1994;120: 817–820.

2 Ezzat S, Asa SL, Couldwell WT, Barr CE, Dodge WE, Vance ML, McCutcheon IE: The prevalence of pituitary adenomas: a systematic review. Cancer 2004;101:613–619.

3 Daly AF, Rixhon M, Adam C, Dempegioti A, Tichomirowa MA, Beckers A: High prevalence of pituitary adenomas: a cross-sectional study in the province of Liege, Belgium. J Clin Endocrinol Metab 2006;91:4769–4775.

4 Agarwal SK, Ozawa A, Mateo CM, Marx SJ: The MEN1 gene and pituitary tumours. Horm Res 2009; 71:131–138.

5 Boikos SA, Stratakis CA: Pituitary pathology in patients with Carney Complex: growth-hormone producing hyperplasia or tumors and their association with other abnormalities. Pituitary 2006;9:203–209.

6 Daly AF, Jaffrain-Rea ML, Ciccarelli A, Valdes-Socin H, Rohmer V, Tamburrano G, Borson-Chazot C, Estour B, Ciccarelli E, Brue T, Ferolla P, Emy P, Colao A, De Menis E, Lecomte P, Penfornis F, Delemer B, Bertherat J, Wémeau JL, De Herder W, Archambeaud F, Stevenaert A, Calender A, Murat A, Cavagnini F, Beckers A: Clinical characterization of familial isolated pituitary adenomas. J Clin Endocrinol Metab 2006;91:3316–3323.

7 Soares BS, Frohman LA: Isolated familial somatotropinoma. Pituitary 2004;7:95–101.

8 Chandrasekharappa SC, Guru SC, Manickam P, Olufemi SE, Collins FS, Emmert-Buck MR, Debelenko LV, Zhuang Z, Lubensky IA, Liotta LA, Crabtree JS, Wang Y, Roe BA, Weisemann J, Boguski MS, Agarwal SK, Kester MB, Kim YS, Heppner C, Dong Q, Spiegel AM, Burns AL, Marx SJ: Positional cloning of the gene for multiple endocrine neoplasia-type 1. Science 1997;276:404–407.

9 Kirschner LS, Carney JA, Pack SD, Taymans SE, Giatzakis C, Cho YS, Cho-Chung YS, Stratakis CA: Mutations of the gene encoding the protein kinase A type I-alpha regulatory subunit in patients with the Carney complex. Nat Genet 2000;26:89–92.

10 Vierimaa O, Georgitsi M, Lehtonen R, Vahteristo P, Kokko A, Raitila A, Tuppurainen K, Ebeling TM, Salmela PI, Paschke R, Gündogdu S, De Menis E, Mäkinen MJ, Launonen V, Karhu A, Aaltonen LA: Pituitary adenoma predisposition caused by germline mutations in the AIP gene. Science 2006;312:1228–1230.

11 Daly AF, Vanbellinghen JF, Khoo SK, et al: Aryl hydrocarbon receptor-interacting protein gene mutations in familial isolated pituitary adenomas: analysis in 73 families. J Clin Endocrinol Metab 2007;92:1891–1896.

12 Toledo RA, Lourenco DM Jr, Liberman B, Cunha-Neto MB, Cavalcanti MG, Moyses CB, Toledo SPA, Dahia PL: Germline mutation in the aryl hydrocarbon receptor interacting protein gene in familial somatotropinoma. J Clin Endocrinol Metab 2007;92:1934–1937.

13 Naves LA, Daly AF, Vanbellinghen JF, Casulari LA, Spilioti C, Magalhães AV, Azevedo MF, Giacomini LA, Nascimento PP, Nunes RO, Rosa JW, Jaffrain-Rea ML, Bours V, Beckers A: Variable pathological and clinical features of a large Brazilian family harboring a mutation in the aryl hydrocarbon receptor-interacting protein gene. Eur J Endocrinol 2007;157:383–391.

14 Yamada S, Yoshimoto K, Sano T, Takada K, Itakura M, Usui M, Teramoto A: Inactivation of the tumor suppressor gene on 11q13 in brothers with familial acrogigantism without multiple endocrine neoplasia type 1. J Clin Endocrinol Metab 1997;82:239–242.

15 Gadelha MR, Prezant TR, Une KN, Glick RP, Moskal SF 2nd, Vaisman M, Melmed S, Kineman RD, Frohman LA: Loss of heterozygosity on chromosome 11q13 in two families with acromegaly/gigantism is independent of mutations of the multiple endocrine neoplasia type I gene. J Clin Endocrinol Metab 1999;84:249–256.

16 Gadelha MR, Une KN, Rohde K, Vaisman M, Kineman RD, Frohman LA: Isolated familial somatotropinomas: establishment of linkage to chromosome 11q13.1–11q13.3 and evidence for a potential second locus at chromosome 2p16–12. J Clin Endocrinol Metab 2000;85:707–714.

17 Luccio-Camelo DC, Une KN, Ferreira RE, Khoo SK, Nickolov R, Bronstein MD, Vaisman M, Teh BT, Frohman LA, Mendonça BB, Gadelha MR: A meiotic recombination in a new isolated familial somatotropinoma kindred. Eur J Endocrinol 2004;150:643–648.

18 Soares BS, Eguchi K, Frohman LA: Tumor deletion mapping on chromosome 11q13 in eight families with isolated familial somatotropinoma and in 15 sporadic somatotropinomas. J Clin Endocrinol Metab 2005;90:6580–6587.

19 Teh BT, Kytölä S, Farnebo F, et al: Mutation analysis of the MEN1 gene in multiple endocrine neoplasia type 1, familial acromegaly and familial isolated hyperparathyroidism. J Clin Endocrinol Metab 1998;83:2621–2626.

20 Jorge BH, Agarwal SK, Lando VS, Salvatori R, Barbero RR, Abelin N, Levine MA, Marx SJ, Toledo SP: Study of the multiple endocrine neoplasia type 1, growth hormone-releasing hormone receptor, Gs alpha, and Gi2 alpha genes in isolated familial acromegaly. J Clin Endocrinol Metab 2001;86:542–544.

21 De Menis E, Prezant TR: Isolated familial somatotropinomas: clinical features and analysis of the MEN1 gene. Pituitary 2002;5:11–15.

22 Carver LA, LaPres JJ, Jain S, Dunham EE, Bradfield CA: Characterization of the Ah receptor-associated protein, ARA9. J Biol Chem 1998;273:33580–33587.

23 Meyer BK, Petrulis JR, Perdew GH: Aryl hydrocarbon (Ah) receptor levels are selectively modulated by hsp90-associated immunophilin homolog XAP2. Cell Stress Chaperones 2000;5:243–254.

24 Karhu A, Aaltonen LA: Susceptibility to pituitary neoplasia related to MEN-1, CDKN1B and AIP mutations: an update. Hum Mol Genet 2007;spec No 1:R73 9.

25 Bolger GB, Peden AH, Steele MR, MacKenzie C, McEwan DG, Wallace DA, Huston E, Baillie GS, Houslay MD: Attenuation of the activity of the cAMPspecific phosphodiesterase PDE4A5 by interaction with the immunophilin XAP2. J Biol Chem 2003;278:3351–3363.

26 de Oliveira SK, Hoffmeister M, Gambaryan S, Muller-Esterl W, Guimaraes JA, Smolenski AP: Phosphodiesterase 2A forms a complex with the cochaperone XAP2 and regulates nuclear translocation of the aryl hydrocarbon receptor. J Biol Chem 2007;282:13656–13663.

27 Leontiou CA, Gueorguiev M, van der Spuy J, et al: The role of the aryl hydrocarbon receptor-interacting protein gene in familial and sporadic pituitary adenomas. J Clin Endocrinol Metab 2008;93:2390–2401.

28 Daly AF, Tichomirow MA, Beckers A: Update on familial pituitary tumors: from multiple endocrine neoplasia type 1 to familial isolated pituitary adenoma. Horm Res 2009;71(suppl 1):105–111.

29 Daly AF, Tichomirowa MA, Beckers A: Genetic, molecular and clinical features of familial isolated pituitary adenomas. Horm Res 2009;71(suppl 2):116–122.

30 Cazabat L, Guillaud-Bataille M, Bertherat J, Raffin-Sanson ML: Mutations of the gene for the aryl hydrocarbon receptor-interacting protein in pituitary adenomas. Horm Res 2009;71:132–141.

31 Vergès B, Boureille F, Goudet P, Murat A, Beckers A, Sassolas G, Cougard P, Chambe B, Montvernay C, Calender A: Pituitary disease in MEN type 1 (MEN1): data from the France-Belgium MEN1 multicenter study. J Clin Endocrinol Metab 2002;87:457–465.

32 Piotrowska K, Pellegata NS, Rosemann M, Fritz A, Graw J, Atkinson MJ: Mapping of a novel MEN-like syndrome locus to rat chromosome 4. Mamm Genome 2004;15:135–141.

33 Pellegata NS, Quintanilla-Martinez L, Siggelkow H, Samson E, Bink K, Höfler H, Fend F, Graw J, Atkinson MJ: Germ-line mutations in p27Kip1 cause a multiple endocrine neoplasia syndrome in rats and humans. Proc Natl Acad Sci USA 2006;103:15558–15563.

34 Georgitsi M, Raitila A, Karhu A, van der Luijt RB, Aalfs CM, Sane T, Vierimaa O, Mäkinen MJ, Tuppurainen K, Paschke R, Gimm O, Koch CA, Gündogdu S, Lucassen A, Tischkowitz M, Izatt L, Aylwin S, Bano G, Hodgson S, De Menis E, Launonen V, Vahteristo P, Aaltonen LA: Germline CDKN1B/p27Kip1 mutation in multiple endocrine neoplasia. J Clin Endocrinol Metab 2007;92:3321–3325.

35 Agarwal SK, Mateo CM, Marx SJ: Rare germline mutations in cyclin-dependent kinase inhibitor genes in multiple endocrine neoplasia type 1 and related states. J Clin Endocrinol Metab 2009;94:1826–1834.

36 Tichomirowa MA, Daly AF, Pujol J, et al: An analysis of the role of cyclin dependent kinase inhibitor 1B (CDKN1B) gene mutations in 86 families with familial isolated pituitary adenomas (FIPA). Ann Meet Endocrine Soc, Washington, 2009, poster P2–118.

37 Stratakis CA, Carney JA, Lin JP, Papanicolaou DA, Karl M, Kastner DL, Pras E, Chrousos GP: Carney complex, a familial multiple neoplasia and lentiginosis syndrome: analysis of 11 kindreds and linkage to the short arm of chromosome 2. J Clin Invest 1996;97:699–705.

38 Matyakhina L, Pack S, Kirschner LS, Pak E, Mannan P, Jaikumar J, Taymans SE, Sandrini F, Carney JA, Stratakis CA: Chromosome 2 (2p16) abnormalities in Carney complex tumours. J Med Genet 2003;40:268–277.

39 Stratakis CA, Kirschner LS: Isolated familial somatotropinomas: does the disease map to 11q13 or to 2p16? J Clin Endocrinol Metab 2000;85:4920–4921.

40 Khoo SK, Pendek R, Nickolov R, Luccio-Camelo DC, Newton TL, Massie A, Petillo D, Menon J, Cameron D, Teh BT, Chan SP: Genome-wide scan identifies novel modifier loci of acromegalic phenotypes for isolated familial somatotropinoma. Endocr Relat Cancer 2009;16:1057–1063.

41 Filopanti M, Ronchi C, Ballarè E, Bondioni S, Lania AG, Losa M, Gelmini S, Peri A, Orlando C, Beck-Peccoz P, Spada A: Analysis of somatostatin receptors 2 and 5 polymorphisms in patients with acromegaly. J Clin Endocrinol Metab 2005;90:4824–4828.

Rodrigo A. Toledo, PhD
Professor Sergio P.A. Toledo
Av. Dr. Arnaldo, 455–5º andar Cerqueira Cesar
012406–903 São Paulo (Brazil)
Tel. +55 11 3061-7226, Fax +55 11 3061-7252, E-Mail: toledorodrigo@usp.br/toldo@usp.br

Serum Levels of 20K-hGH and 22K-hGH Isoforms in Acromegalic Patients

Giovanna A.B. Lima[a,b] · Mônica R. Gadelha[a,b] · Christian J. Strasburger[c] · Zida Wu[c]

[a]Division of Endocrinology, Department of Internal Medicine, Hospital Universitário Clementino Fraga Filho/Universidade Federal do Rio de Janeiro, and [b]Instituto Estadual de Diabetes e Endocrinologia Luiz Capriglione, Rio de Janeiro, Brazil; [c]Division of Clinical Endocrinology, Department of Medicine, Charité Universitätsmedizin, Berlin, Germany

Abstract

The diagnosis and treatment decisions in acromegaly depend on the measurement of growth hormone (GH) and insulin-like growth factor I (IGF-I) levels. The occurrence of different GH isoforms in the serum, mainly 22K-hGH and 20K-hGH, is a source of heterogeneity of GH results measured by different immunoassays. Since it has been previously reported that the proportion of 20K-hGH is increased in patients with active acromegaly, it might be useful to know the GH isoforms' pattern not to underdiagnose or undertreat acromegalic patients.

Copyright © 2010 S. Karger AG, Basel

Introduction

The diagnosis of acromegaly requires biochemical confirmation of high growth hormone (GH) and insulin-like growth factor I (IGF-I) levels. The same parameters are used for therapeutic decisions. Therefore, we must be aware that there are many sources of heterogeneity of GH results measured by different immunoassays, such as: the assay calibrator, the epitope specificity of antibodies used, the assay design (competitive vs. sandwich), susceptibility to interference by endogenous hGHBP and heterogeneity of human GH (hGH) isoforms circulating in serum.

Growth hormone exists as a mixture of multiple molecular isoforms arising from posttranscriptional and posttranslational modifications. The major isoform is the 22-kDa hGH (22K-hGH), which accounts for 60–70% of all pituitary GH molecules and is composed by 191 amino acid residues, whereas 20-kDa hGH (20K-hGH) is the second most abundant isoform and represents 10% of total GH in the pituitary. GH

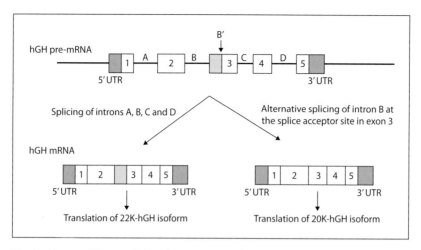

Fig. 1. Human GH pre-mRNA splicing. Boxes indicate the mRNA coding regions (exons 1, 2, 3, 4 and 5) and intervening sequences are thick lines (introns A, B, C and D). B' indicates the position of alternative splicing, within exon 3. As shown in the left, the regular splicing of the pre-mRNA produces an mRNA that is translated to produce the 22K-hGH precursor. In the right, alternative splicing gives rise to a variant GH molecule: the 20K-hGH isoform, that lacks 15 amino acids from the beginning of exon 3. GH = Growth hormone; mRNA = messenger RNA; UTR = untranslated region.

fragments/segments, such as the GH 1–43 (5K-hGH) and GH 44–191 (17K-hGH), also exist and have different molecular sizes [1].

The 20K-hGH Isoform – Structure and Biological Function

Transcripts of the hGH gene (GH-N) undergo two different types of splicing: (a) the regular splicing between intron B and exon 3 that originates the 22K-hGH isoform mRNA, and (b) an alternative splicing within exon 3 that produces the 20K-hGH isoform [2, 3]. The regular splicing is preferentially used, but both isoforms are produced in the pituitary (fig. 1).

The 20K-hGH is composed of 176 amino acids and differs from the 22K-hGH by an internal deletion of a 15 amino acid sequence that includes residues 32–46 of 22K-hGH [2–4]. Initially, the 20K-hGH was believed to lack biological activity due to the altered receptor binding site 1 conformation [5]. However, although the 20K-hGH has low binding affinity to the human GH receptor [6], subsequent studies have shown agonistic actions comparable to those of the 22kD-hGH [2, 7–11]. Nonetheless, the antinatriuretic [11, 12], diabetogenic [11, 13] and lactogenic [14] activities are lower than that of the 22K-hGH (table 1).

The 20K-hGH appears to be co-secreted with 22K-hGH in a fixed ratio and no specific stimulus for preferential secretion of GH isoforms has been identified [15].

Table 1. Biological effects of the 20K-hGH isoform

Effect	Model	Potency
Linear growth	hypophysectomized rats [2]	= 22K-hGH
	transgenic mice [7]	= 22K-hGH
	dwarf rats [8]	= 22K-hGH
Lipolysis	rat adipocytes [9]	= 22K-hGH
	humans (normal) [10]	= 22K-hGH
	humans (adult GHD) [11]	= 22K-hGH
Bone turn over	humans (adult GHD) [11]	= 22K-hGH
Lactogenic activity (hPRLR)	BaF/B03-hPRLR and CHO-hPRLR cells [14]	<22K-hGH
Antinatriuretic	rats [12]	<22K-hGH
	humans (adult GHD) [11]	= 22K-hGH
Diabetogenic	rats [13]	<22K-hGH
	humans (adult GHD) [11]	≤22K-hGH

GHD = Growth hormone deficiency; BaF/B03 cells = bone marrow-derived murine IL-3-dependent pro-B cell line; hPRLR = human prolactin receptor; CHO cells = Chinese hamster ovarian cells.

The metabolic clearance of 20K-hGH is slower than that of 22K-hGH, partially due to its tendency to dimerize [16, 17].

Laboratory Measurement of Growth Hormone Isoforms

The Non-22K-hGH Assay

Because of the difficulty to develop assays completely specific for the minor GH isoforms, mainly the 20K-hGH, a 'non-22K-hGH assay' was designed [18]. This 22K-hGH exclusion assay made it possible to evaluate the proportion of non-22K-hGH in human blood under different physiological and pathophysiological conditions [19–22]. In this assay, monomeric and dimeric 22K-hGH are removed from the serum by a specific anti-22K-hGH antibody and by paramagnetic beads coated with a second antibody. After 22K-hGH extraction from the serum, non-22K-hGH levels are measured by a polyclonal antibody-based immunoradiometric assay. Simultaneously, total GH levels are measured in another aliquot of serum, allowing the determination of the proportion of non-22K-hGH isoforms to total GH in the sample.

Assays for the 20K-hGH Isoform

It has been very difficult to develop a specific anti-20K-hGH antibody which does not cross-react with 22K-hGH. Because 22K-hGH is the predominant GH isoform in the serum, even a small degree of cross-reactivity with the 22K-hGH presents a significant barrier to the accurate measurement of the 20K-hGH. Since the only structural difference between the 20K- and 22K-hGH isoform is the deletion of amino acid residues 32–46, the specific antibody should recognize the new amino acid sequence F31-N32 of 20K-hGH (they are F31 and N47 in 22K-hGH), or the slight conformational alteration caused by this deletion.

In the last two decades, three 20K-hGH specific monoclonal antibodies (mAb) have been described [23–25] and the serum concentrations of 20K-hGH have been investigated in different clinical settings.

Mellado et al. [23] reported for the first time the generation of an anti-20K-specific antibody (hGH-33) using the native protein as an immunogen. They suggested that this antibody is able to recognize a conformational epitope present only in the 20K-hGH isoform because it does not recognize peptides containing the new amino acid sequence F31-N32 or the 20K-hGH isoform under circumstances that may alter its three dimensional structure. Using this mAb, they developed sandwich enzyme immunoassay (EIA) for 20K-hGH. However, due to its sensitivity limitation (4 ng/ml), it is not appropriate for assessing the serum levels of 20K-hGH isoform under physiological conditions.

Hashimoto et al. [24] generated the mAb D05, a specific mAb to the recombinant 20K-hGH isoform. Since mAb D05 did not stain the 20K-hGH isoform in Western blot analysis, it seems to recognize a conformational epitope present only in this isoform, which may disappear upon denaturation. The combination of mAb D05 bound to solid-phase and affinity-purified anti-20K-hGH polyclonal antibody as the labeled antibody led to the construction of a highly specific sandwich EIA, with sensitivity of 10 pg/ml.

Recently, we (Z.W. and C.J.S.) developed specific mAbs for recombinant 20K- and 22K-hGH isoforms. Using pairs of specific mAbs for the 20K-hGH and 22K-hGH, we constructed highly sensitive time-resolved fluorescence immunoassays (TR-IFMA) for each hGH isoform [25]. The working range of both 20K-hGH and 22K-hGH specific assays (0.02–20 and 0.02–50 ng/ml, respectively) is appropriate for evaluating the 20K- and 22K-hGH concentrations in most human serum samples.

Assays for the 22K-hGH Isoform

Several assays have been designed to measure only the 22K-hGH [26, 27]. In fact, as the 20K-hGH corresponds to about 10% of total GH, measuring only the 22K-hGH isoform could underestimate the total pituitary GH production and cause underdiagnosis or undertreatment of acromegaly.

Growth Hormone Isoforms in Acromegaly

Several studies evaluated the GH isoforms (20K-hGH and 22K-hGH) in acromegalic patients.

Baumann et al. [28], via immunoadsorbent chromatography, demonstrated that the monomeric GH isoforms excreted in the urine of an acromegalic patient are qualitatively indistinguishable from those in normal individuals.

Boguszewski et al. [19], using the 22K-hGH exclusion assay, studied 15 acromegalic men before and one year after trans-sphenoidal surgery. They found that untreated acromegalic patients have an increased proportion of circulating non-22K-hGH when compared with controls (26.6 vs. 17.4%), and the values correlated significantly to the tumor size, mean 24-hour GH concentration and serum prolactin levels. After surgery, patients not truly cured, with mean 24-hour GH concentration of 1 µg/l or more, had an increased proportion of non-22K-hGH when compared to those with GH levels <1 µ/l. In the group of patients not truly cured, the proportion of non-22K-hGH was similar to that in untreated patients (34 vs. 26.6%, respectively). On the other hand, in the group of cured patients, the proportion of non-22K-hGH was comparable to those in the controls (15.2 vs. 17.4%, respectively).

Tsushima et al. [29], through a sensitive enzyme-linked immunosorbent assay specific to the 20K- hGH and 22K-hGH isoforms, verified that the %20K-hGH (ratio of 20K to 20K plus 22K) is significantly increased in patients with active acromegaly when compared to normal subjects (9.2 vs. 6.3%, respectively). In contrast, the %20K-hGH in successfully treated patients did not differ from that in normal subjects.

Only one study evaluated the serum levels of 20K-hGH and 22K-hGH in acromegalic patients treated with subcutaneous octreotide, and demonstrated a significant reduction in the %20K-hGH in 2/4 patients treated with this drug [30].

In our own experience, using the TR-IFMA specific for the 20K- and 22K-hGH isoforms [25], acromegalic patients with active disease present a significantly higher %20K-hGH in comparison to matched healthy controls [31].

Conclusions

GH circulates as a mixture of several molecular isoforms and fragments which differ in size, immunoreactivity and bioactivity. This concept of GH heterogeneity must be taken into account when measuring GH levels for diagnostic and therapeutic purposes in GH-dependent disorders, such as acromegaly. Because there are only a few reports on the role of different GH isoforms in acromegaly, further studies are necessary to elucidate whether the measurement of these isoforms by specific immunoassays will have a place in the biochemical evaluation of this disease.

References

1 Baumann G: Growth hormone heterogeneity: genes, isoforms, variants and binding proteins. Endocr Rev 1991;12:424.
2 Lewis UJ, Dunn JT, Bonewald LF, Seavey BK, VanderLaan WP: A naturally occurring variant of human growth hormone. J Biol Chem 1978;253: 2679.
3 DeNoto FM, Moore DD, Goodman HM: Human growth hormone sequence and DNA structure: possible alternative splicing. Nucleic Acids Res 1981;9: 3719.
4 Lewis UJ, Bonewald LF, Lewis LJ: The 20,000-Dalton variant of human growth hormone: location of the amino acid deletions. Biochem Biophys Res Commun 1980;92:511.
5 McCarter J, Shaw MA, Winer LA, Baumann G: The 20,000 Da variant of human growth hormone does not bind to growth hormone receptors in human liver. Mol Cell Endocrinol 1990;73:11.
6 Uchida H, Banba S, Wada M, Matsumoto K, Ikeda M, Naito N, Tanaka E, Honjo M: Analysis of binding properties between 20 kDa human growth hormone (hGH) and hGH receptor (hGHR): the binding affinity for hGHR extracellular domain and mode of receptor dimerization. J Mol Endocrinol 1999;23:347.
7 Stewart TA, Clift S, Pitts-Meek S, Martin L, Terrell TG, Liggitt D, Oakley H: An evaluation of the functions of the 22-kilodalton (kDa), the 20-kDa, and the N-terminal polypeptide forms of human growth hormone using transgenic mice. Endocrinology 1992;130:405.
8 Ishikawa M, Tachibana T, Kamioka T, Horikawa R, Katsumata N, Tanaka T: Comparison of the somatogenic actions of 20kDa and 22kDa human growth hormone in spontaneous dwarf rats. GH IGF Res 2000;10:199.
9 Uchida H, Naito N, Asada N, Wada M, Ikeda M, Kobayashi H, Asanagi M, Mori K, Fujita Y, Konda K, Kusuhara N, Kamioka T, Nakashima K, Honjo M: Secretion of authentic 20-kDa human growth hormone (20K hGH) in *Escherichia coli* and properties of the purified product. J Biotechnol 1997;55: 101.
10 Asada N, Takahashi Y, Wada M, Naito N, Uchida H, Ikeda M, Honjo M: GH induced lipolysis stimulation in 3T3-L1 adipocytes stably expressing hGHR: analysis on signaling pathway and activity of 20K hGH. Mol Cell Endocrinol 2000;162:121.
11 Hayakawa M, Shimazaki Y, Tsushima T, Kato Y, Takano K, Chihara K, Shimatsu A, Irie M: Metabolic effects of 20-kilodalton human growth hormone (20K-hGH) for adults with growth hormone deficiency: results of an exploratory uncontrolled multicenter clinical trial of 20K-hGH. J Clin Endocrinol Metab 2004;89:1562.
12 Satozawa N, Takezawa K, Miwa T, Takahashi S, Hayakawa M, Ooka H: Differences in the effects of 20 K- and 22 K-hGH on water retention in rats. GH & IGF Res 2000;10:187.
13 Takahashi S, Shiga Y, Satozawa N, Hayakawa M: Diabetogenic activity of 20 kDa human growth hormone (20K-hGH) and the 22K-hGH in rats. GH & IGF Res 2001;11:110.
14 Tsunekawa B, Wada M, Ikeda M, Uchida H, Naito N, Honjo M: The 20-kilodalton (kDa) human growth hormone (hGH) differs from the 22-kDa hGH in the effect on the human prolactin receptor. Endocrinology 1999;140:3909.
15 Baumann G, Stolar MW, Amburn K: Molecular forms of circulating growth hormone during spontaneous secretory episodes and in the basal state. J Clin Endoc Metab 1985;60:1216.
16 Baumann G, Stolar MW, Buchanan TA: Slow metabolic clearance rate of the 10,000-Dalton variant of human growth hormone: implications for biological activity. Endocrinology 1985;117:1309.
17 Leung K, Howe C, Gui LYY, Trout G, Veldhuis JD, Ho KKY: Physiological and pharmacological regulation of 20-kDa growth hormone. Am J Physiol Endocrinol Metab 2002;283:E836.
18 Boguszewski CL, Hynsjö L, Johannsson G, Gengston BA, Carlsson LM: 22-kD growth hormone exclusion assay: a new approach to measurement of non-22-kD growth hormone isoforms in human blood. Eur J Endocrinol 1996;135:573.
19 Boguszewski CL, Johannsson G, Bengtsson B, Johansson A, Carlsson B, Carlsson LMS: Circulating non-22-kilodalton growth hormone isoforms in acromegalic men before and after transsphenoidal surgery. J Clin Endocrinol Metab 1997;82:1516.
20 Boguszewski CL, Jansson C, Boguszewski MC, Rosberg S, Carlsson B, Albertsson-Wickland K, Carlsson LM: Increased proportion of circulating non-22-kilodalton growth hormone isoforms in short children: a possible mechanism for growth failure. J Clin Endocrinol Metab 1997;82:2944.
21 Boguszewski CL, Boguszewski MC, de Zegher F, Carlsson B, Carlsson LM: Growth hormone isoforms in newborns and postpartum women. Eur J Endocrinol 2000;142:353.

22 Wallace JD, Cueno RC, Bidlingmaier M, Lundberg PA, Carlsson L, Boguszewski CL, Hay J, Healy ML, Napoli R, Dall R, Rosen T, Strasburger CJ: The response of molecular isoforms of growth hormone to acute exercise in trained adult males. J Clin Endocrinol Metab 2001;86:200.

23 Mellado M, Rodriguez-Frade JM, Kremer L, Martinez-Alonso C: Characterization of monoclonal antibodies specific for the human growth hormone 22K and 20K isoforms. J Clin Endocrinol Metab 1996;81:1613.

24 Hashimoto Y, Ikeda I, Ikeda M, Takahashi Y, Hosaka M, Uchida H, Kono N, Fukui H, Makino T, Honjo M: Construction of a specific and sensitive sandwich enzyme immunoassay for 20 kDa human growth hormone. J Immunol Methods 1998;221:77.

25 Wu Z, Bidlingmaier M, Keller A, Keller E, Strasburger CJ: Development of ultra sensitive human growth hormone (hGH) isoform specific imunnoassays and their use to detect doping with recombinant hGH. Proc 85th Meeting Endocr Soc, Philadelphia, 2003, pp OR32–OR33.

26 Celniker AC, Chen AB, Wert RM Jr, Sherman BM: Variability in the quantitation of circulating growth hormone using commercial immunoassays. J Clin Endocrinol Metab 1989;68:469.

27 Vieira JGH, Ando MHK, Nishida SK, Macedo CD, Abucham JZ: Clinical utility of a 22-kDa growth hormone-specific assay. Braz J Med Biol Res 1992; 25:243.

28 Baumann G, MacCart JG, Amburn K: The molecular nature of circulating growth hormone in normal and acromegalic man: evidence for a principal minor monomeric forms. J Clin Endocrinol Metab 1983;56:946.

29 Tsushima T, Katoh Y, Miyachi Y, Chihara K, Teramoto A, Irie M, Hashimoto Y: Serum concentrations of 20K human growth hormone (20K-hGH) measured by a specific enzyme-linked immunosorbent assay. J Clin Endocrinol Metab 1999;84:317.

30 Murakami Y, Shimizu T, Yamamoto M, Kato Y: Serum levels of 20 kilodalton human growth hormone (20K-hGH) in patients with acromegaly before and after treatment with octreotide and transsphenoidal surgery. Endocr J 2004;51:343.

31 Lima GAB, Wu Z, Silva CM, Barbosa FR, Dias JS, Schrank Y, Strasburger CJ, Gadelha MR: Growth hormone isoforms in acromegalic patients before and after treatment with octreotide LAR. GH IGF Res 2009;in press.

Giovanna A.B. Lima
Hospital Universitário Clementino Fraga Filho
Rua Professor Rodolpho Paulo Rocco, 255, 9E23, Cidade Universitária
Ilha do Fundão – RJ 21941–913 (Brasil)
Tel. +55 21 2562 2323 and +55 21 9241 1533, Fax +55 21 2562 2111, E-Mail gibalarini@gmail.com

Pituitary Carcinomas

Annamaria Colao[a] · Arantzazu Sebastian Ochoa[b] ·
Renata Simona Auriemma[a] · Antongiulio Faggiano[a] ·
Rosario Pivonello[a] · Gaetano Lombardi[a]

[a]Departments of Molecular and Clinical Endocrinology and Oncology 'Federico II' University of Naples, Naples, Italy; [b]Department of Endocrinology, University Hospital San Cecilio, Granada, Spain

Abstract

Pituitary carcinoma is a extremely rare and is characterized by a very poor prognosis. Even if at diagnosis the presence of metastases is required to define a pituitary carcinoma, the lesion was almost invariably diagnosed first as a benign pituitary tumor, that after a variable period of latency, ranging from few months to many years, changed its natural course to an aggressive pituitary tumor poorly responsive to therapy. Recent studies have partially clarified its molecular pathogenesis and found possible markers of aggressiveness in order to make an earlier diagnosis, when still treatment could improve their prognosis. Most pituitary carcinomas are functioning, and ACTH- and PRL-secreting carcinomas are the most frequent. Treatment includes surgery, radiotherapy, medical therapy and chemotherapy, but the poor results with current therapies should prompt all investigators to better understand its pathogenesis and searching new molecular targets for treatments.

Copyright © 2010 S. Karger AG, Basel

Pituitary tumors represent nearly 15% of intracranial tumors and the third most common intracranial tumor after meningiomas and gliomas [1, 2]. The great majority of pituitary tumors are noninvasive benign adenomas that either remain within the sella or may exhibit expansive growth to surrounding tissues [3]. Invasive adenomas are a group of pituitary tumors with biological behavior between pituitary adenomas and carcinomas [4–10]. Pituitary carcinomas are classically defined as pituitary tumors with subcranial, brain or systemic metastases [1, 2, 4–8]. In general, pituitary carcinomas are associated with poor prognosis as therapeutic options are limited [1, 2, 4–8, 9, 11]. Due to their relative rarity, many unanswered questions regarding pathophysiology have been raised. In particular, it is still unclear if pituitary carcinomas develop from adenomas or they occur de novo and if hormonal subtype does affect tumor aggressiveness, treatment outcome and prognosis. Additionally, prediction of tumor invasiveness by studying proliferation markers, onco-protein expression, and DNA

Table 1. Reported cases of pituitary carcinomas

Tumor type	Kaltsas et al. [1] up to 1998	Ragel et al. [5] up to 2004	Since 2004
ACTH-secreting tumors	24	59	2 [12, 21]
PRL-secreting tumors	24	46	7 [13–16,18–20]
GH-secreting tumors	12	9	0
Nonfunctioning tumors	29	17	0
FSH-LH-secreting tumors	1	7	1 [13]
TSH-secreting tumors	1	1	1 [15]

index is still incomplete. A perspective in research is to identify specific oncogenic mutations associated with disease progression for future gene therapy.

This review aims at summarizing the most relevant aspects of epidemiology, pathogenesis, diagnosis and treatment of pituitary carcinomas.

Epidemiology

Approximately 140 cases of pituitary carcinoma were reported in an extensive review of the British literature up to 2005 [6], and additional 10 cases have been reported afterwards [1, 2, 4–8, 12–21] (table 1). Pituitary carcinomas have no gender difference and are diagnosed at a mean age of 44 years, with a latency period of 7 years after the initial diagnosis of a pituitary tumor [1, 6]. Approximately 80% of patients died within 8 years and 66% died within 1 year after presentation of the metastases [6].

The etiology of pituitary carcinomas remains unclear; however a carcinoma is considered to be induced by: (1) previous pituitary tumor irradiation; (2) microscopic tumor seeding from a previous pituitary surgery, and (3) malignant progression of a pituitary adenoma (alternatively a de novo carcinoma) (table 2). The cumulative incidence of second brain tumors was 2.4% at 20 years, with a relative risk of 10.5 compared with the general population [22–26]. Tumor seeding from opening of the subarachnoid space during previous pituitary surgery has also been suggested, but no definitive correlation has been made yet [1, 5, 6, 22, 23, 27]. Progression from macroadenomas to carcinoma has also been considered [1, 6, 28, 29]. The monoclonal origin of primitive pituitary tumors and secondary metastases, similar histological findings and molecular markers, and progressive accumulation of genetic aberrations in oncogenes and/or tumor suppressor genes [30, 31], suggest that pituitary carcinomas arise after transformation of initially benign pituitary tumors [1, 4, 9, 32–34].

Table 2. Hypotheses of pituitary carcinoma etiology

Hypotheses	Mechanism and evidence	References
Radiation	a consequence of previous irradiation in the treatment of pituitary tumor	19, 21, 22–25, 74
Surgery	microscopic tumor seeding from a previous pituitary surgery	1, 5, 6, 21, 22, 26
Adenoma to carcinoma	most initial presentation as macroadenomas long time interval for progression monoclonal origin progressive accumulation of aberrations in oncogenes and/or tumor suppressor genes similar histological findings similar molecular markers	1, 4, 6, 27, 28
De novo	separate clonal expansion	31

Table 3. Molecular pathogenesis pituitary carcinomas

Chromosomal aberrations	Oncogenes	Tumor suppressor gene
Gains of chromosome 5, 7p and 14q	Gsp	p53 retinoblastoma
Loss of chromosome 1p, 3p, 10q, 11q and 22q	H-ras oncogene	nm23 p27

Only one case reported presence of distinct clones in the primitive and metastatic lesions [32]. One recent hypothesis is disruption of the pituitary tumor transforming gene and p21 pathway that induce cell senescence and thus prevent transformation of pituitary adenomas into carcinomas [35].

Pathogenesis

The genetic basis of pituitary tumors is largely unknown but chromosomal aberrations and/or accumulation of genetic abnormalities seem to contribute to the development of pituitary tumors and, mainly pituitary carcinomas (table 3).

Chromosomal aberrations are more frequently discovered in metastases than in primitive pituitary carcinomas. An average of 8.3 chromosomal imbalances per tumor has been demonstrated in metastatic pituitary carcinomas: the most common are gains of chromosome 5, 7p, 13q and 14q and loss of chromosomes 1p, 3p, 10q, 11q and 22q [36, 37].

Oncogenes

The GSP is the oncogen which has been most significantly associated with pituitary tumors, but GSP-activating mutations are rarely detected in tumor other than GH-secreting adenomas [1, 31, 38, 39], suggesting that this gene is more important in pituitary tumorigenesis than in development of carcinomas. The H-RAS gene encodes for a GTP-binding protein that is important in cellular proliferation and differentiation and particularly involved in early stages of tumorigenesis. Mutational activation of H-RAS, which plays a critical role in carcinogenesis, has been observed in metastases from pituitary carcinoma but not in their respective primitive pituitary tumors [40]. This suggests that point mutations in H-RAS are not associated with pituitary tumorigenesis, but may be important in the formation and growth of pituitary metastases [41].

Tumor Suppressor Genes

The *p53* gene is a tumor suppressor gene located on chromosome 17p13.1, which encodes a nuclear phosphoprotein that plays an essential role in the control of cell proliferation. Although p53 mutations were found in pituitary adenomas or carcinomas and their respective metastases, they were not detected in exons 5–8, where they usually occur. Thus, p53 probably has little or no role in pituitary tumorigenesis [9].

The retinoblastoma (Rb) gene encodes for a protein regulating cell cycle with an important role in controlling cell differentiation and survival. The Rb gene mutation has not been detected in human pituitary tumors [41] but a deletion of gene locus next to Rb gene was detected in most aggressive adenomas and in carcinomas.

The *nm23* gene, encodes a purine-binding factor, which prevents the progression of cell cycle and consequently reduces the metastatic potential of the tumor. Reduced expression of nm23 protein has been shown in pituitary carcinomas, where the level of nm23 expression was inversely correlated with cavernous sinus invasion rather than with pituitary tumor size per se [42, 43].

The *p27* gene encodes an inhibitor of the kinase activity of the cyclin-CDK complexes, which plays a central role in controlling cell cycle progression. Mutation of the p27 gene has not been documented in human pituitary tumors; nonetheless, nuclear p27 protein expression is decreased in adenomas as compared with normal pituitary, and is lower in pituitary carcinomas than in both normal pituitary and adenomas [8, 44, 45].

In summary, multiple molecular events may occur to initiate the transformation from benign pituitary adenomas to aggressive adenomas and/or carcinomas. These include early chromosomal mutations and expression of pituitary-specific oncogenes. Subsequent permissive factors may include hypothalamic hormone receptors signals, paracrine growth factors signals, and disordered cell cycle regulation but the exact mechanism remains unclear.

Markers of Tumor Aggressiveness

At histology, the ability to infiltrate dura may not reflect the tendency to recur after transsphenoidal excision [46], but mitotic indices seem to be higher in progressive and/or metastatic lesions than benign tumors. The cell proliferation rate is an important biological parameter related to the aggressive potential of pituitary tumors [9, 48]. Different markers have been hypothesized to indicate the aggressiveness of the pituitary tumors and/or the potentiality to transform in pituitary carcinomas.

Markers of Proliferation
Ki-67: Ki-67 antigen, a proliferation marker selectively expressed during G1, S1, G2, and M phases of cell cycle [47]. The growth fraction of pituitary tumors is low, with indices generally lower than 5%. Virtually all noninvasive adenomas had a growth fraction less than 3% [47, 48]. On the other hand, growth fractions were significantly higher among invasive adenomas, and the highest among pituitary carcinomas, when compared to noninvasive adenomas (4.7, 11.9 and 1.4%, respectively) [49]. Because of wide variability in Ki-67 expression in different studies, it is suggested that this is not a major determinant of carcinogenesis.

PCNA: Proliferating cell nuclear antigen (PCNA) is a protein accumulating in the nucleus during the cell cycle, which has a specificity for the S phase higher than Ki-67, and labels only those cells that passed the important G1/S boundary [50], but overestimates cell proliferation in pituitary tumors [48].

Adhesion Molecules: E-cadherin is an important molecule implicated in tumor invasion processes which functions are mediated by catenins α,β,γ. The expression of β-catenin, are significantly lower in pituitary carcinomas than normal pituitary tissue and adenomas [51]. However, the molecular mechanisms responsible for the loss of cadherin/catenin expression in carcinomas are presently unknown.

Local Growth Factor Expression: In the pituitary, various growth factors are expressed and released into the extracellular matrix. Fibroblast growth factors 2 [52] and 4 [53] (FGF-2 and FGF-4), implicated in angiogenesis, are being studied in progression of pituitary carcinomas. The epidermal growth factor (EGF) receptor (EGFR) and its activated form, phospho-EGFR (P-EGFR), are expressed in normal and neoplastic pituitary tissues. Onguru et al. [54] found EGFR overexpressed in carcinomas more than in adenomas. The expression of HER-2/neu proto-oncogene, encoding for a trans-membrane tyrosine kinase receptor belonging to the EGFR family, has been identified in the metastases of ACTH- and gonadotrophin-secreting pituitary carcinomas [55, 56].

p53: Thapar et al. [48], in a study of 70 pituitary adenomas and 7 pituitary carcinomas, demonstrated that the expression of p53 gene product was present in almost all pituitary carcinomas, and in approximately 15% of invasive adenomas, but not in benign adenomas. Other studies, however, did not detect p53 protein in pituitary adenomas either invasive or noninvasive [43].

DNA Ploidy Status: DNA ploidy status (diploid vs. aneuploid) is a useful prognostic indicator in several human neoplasms but not pituitary adenomas [57, 58]. In fact, Scheithauer et al. [2] did not find any difference in the frequency of aneuploidy between non invasive and invasive tumors.

Additional Markers: Galectin-3 (Gal-3), a galactoside-binding protein, is expressed in PRL- and ACTH-secreting adenomas, and its expression was significantly increased in ACTH-secreting carcinomas compared with adenomas [59]. The expression of cyclooxygenase-2, an enzyme overexpressed in several tumors, which is involved in pituitary tumor angiogenesis, was also found to be higher in carcinomas [60]. Bcl-2 is a cytoplasmatic protein whose expression in cancer cells is thought to inhibit apoptosis; Bcl-2 expression was reportedly greater in pituitary carcinomas [61].

In summary, immunohistochemical staining for the expression of the protein produced by mutated p53 gene and for Ki-67 index of cellular proliferation, have been suggested to be among the most promising prognostic markers. The current definition of pituitary carcinoma requires the demonstration of metastases, Ki-67 >3% and/or p53 >3% in a pituitary adenoma: this suggests aggressive behavior and should alert the clinicians to the possibility of a pituitary carcinoma.

Diagnosis

Due to their poor prognosis, rapid identification of pituitary carcinomas is mandatory. Kaltsas et al. [6] proposed the following criteria: (1) a primary pituitary tumor must be identified by histology; (2) an alternative primary tumor has to be excluded; (3) metastases must be identified, either as discontinuous spread subarachnoid metastatic deposits in the brain or systemic deposits from metastases of carcinomas in different organs, and (4) structural features or marker expressions of metastases should correspond or be similar to those of the pituitary tumor.

Histological studies have so far failed to reliably distinguish invasive pituitary tumors and pituitary carcinomas. Morphological markers are of limited usefulness in pituitary tumors: hypercellularity, nuclear pleomorphism, mitotic activity, necrosis, hemorrhage, and even invasion are not reliable indicators of the malignant nature of the tumor [4, 11].

Current neuroimaging modalities, computed tomography (CT) and especially high resolution gadolinium-enhancement magnetic resonance imaging (MRI), have high sensitivity in detecting pituitary lesions and can also be used to demonstrate metastases [1, 6, 8]. A differential diagnosis between invasive adenoma and carcinoma is not possible with the neuroimaging studies. Radionuclide and functional imaging studies using specific radiotracer have also been employed to identify disease extension and metastases. Metaiodobenzylguanidine (MIBG) and ^{111}In- DTPA-D-phe^1-octreotide [62] were not highly successful, though the latter allowed to diagnose a metastatic GH-secreting carcinoma [63, 64], to reveal additional lesions in an ACTH-secreting carcinoma [65] and to reveal tumor recurrences of metastatic lesions at follow-up [66]. More recently, positron emission tomography (PET) using ^{18}fluoro-deoxyglucose as radiolabeled tracer has revealed unsuspected pituitary macroadenomas and also identified metastases from pituitary carcinomas [67].

The general belief is to consider pituitary adenomas with atypical histological characteristics, local invasiveness or aggressive behavior, as at risk to develop in a pituitary carcinoma.

Clinical Findings

Most of the reported pituitary carcinomas were associated to hormone excess (83%), with 35% of the lesions producing ACTH, 33% PRL, 9% GH, 4% LH and/or FSH and only 1% TSH. Nonfunctioning tumors represent 19% of cases, also including silent ACTH, FSH, LH and rare null-cell pituitary carcinomas (table 1). Early reports of non-functioning pituitary carcinomas were probably due to lack of hormone assays and routine use of immunohistochemistry, and a significant number of PRL-secreting carcinomas have likely been unrecognized [1, 2, 6]. The clinical syndrome related to hormone hypersecretion is usually similar to that encountered in patients with benign tumors and hormone levels did not differentiate pituitary carcinomas from other invasive and/or non invasive macroadenomas [1, 4, 6, 8]. However, dramatically high levels of PRL, ACTH, and/or GH despite apparent surgical removal of the tumor as well as partial or complete resistance to dopamine agonists (DA) and/or somatostatin analogs (SSA) or 'treatment escape' after a period of controlled disease could be consequence of metastases or transformation into carcinoma. Those signs in follow-up of a pituitary adenoma should alert the clinician, and metastases should always be excluded. Diabetes insipidus is not a common feature at initial presentation [1].

Symptoms related to metastases vary according to site of involvement; headache and vision defects are typical of development of invasiveness. Metastases in the central nervous system (CNS) usually involve the frontal lobe, parietal-occipital lobe, cerebellum, brainstem, cerebellopontine angle, spinal cord, cauda equina and subaracnoidal space [1]. A single case of intramedullar metastasis was described [12]. Less commonly involved areas were cranial nerves, olfactory bulb and hippocampus. Distant metastases usually involve liver, lymph nodes, bone, lung, and less commonly heart, kidneys, pancreas, eyes, ear, ovaries, myometrium and muscles. Systemic dissemination is believed to occur by lymphatic and/or hematological route. In cases in which the cavernous sinus has been invaded by the tumor, spread probably occurs via the superior petrosal sinus [1, 8].

Treatment

Treatment options of pituitary carcinomas include surgery, radiotherapy, and medical treatment.

Surgery

The best treatment approach to pituitary carcinoma is early surgery. Few series of pituitary carcinoma describing type of surgery employed have been reported [4, 12] and since most described single cases conclusion on significance of type of surgery is hard. There are no doubts about the importance of surgical resection in relief of compressive symptoms and need of histological diagnosis of pituitary tumor, although it is seldom curative. These tumors are largely invasive, infiltrating adjacent and vital tissues. There seems to be no benefit in prognosis when transcranial rather than trans-sphenoidal surgery is advocated, although the former may be necessary when there is extension to the anterior or posterior fossa or other vital structures [1, 6]. The results of surgery in treating isolated and/or symptomatic metastatic lesions, have shown partial remission [68] as well as total resection [63] in two single cases.

Radiotherapy

Radiotherapy (RT) has traditionally been used for prevention of tumor regrowth in large or partially removed pituitary tumors and for local control of expanding tumors and/or metastatic deposits [1, 6, 8]. RT has shown to be effective in achieving local control in many patients when administered to expanding lesions that were later proven to be carcinomas [1, 6]. However, RT given when pituitary carcinomas were already diagnosed, was successful only in a few patients [1, 6]. In CNS metastases, RT was shown to prevent growth and even induce partial regression in some patients [28, 69, 70]. Therefore, RT may have a palliative role, without changing the poor prognosis of the disease.

Focused RT (stereotatic multiarc RT or gamma-knife) can be applied to a smaller area than the conventional one, with reduced risk of radiation-induced damage to surrounding tissues [1, 71]. Only few cases have been reported: focused RT was given to patients who had already received conventional radiotherapy with development of recurrence or metastases, with little palliative results [31, 71, 72].

Radiolabeled-SSA involve the delivery of a toxic dose of radioactivity using various radiation-delivering agents (111-indium, 90-ittrium, 177-lutecium), specific for tumor tissue [73]. It has been applied to neuroendocrine tumors (NET) with successful outcome in some NET refractory to biotherapy and/or chemotherapy [73]. Although this type of treatment has not been applied yet to malignant pituitary tumors expressing somatostatin receptors, it could become an option to consider [73], but the little experience and the risk of toxicity for surrounding normal brain tissue make the indication still restricted.

The incidence of second brain tumors in patients with pituitary adenomas after radiotherapy has been a matter of debate in the last few years [22, 23]. The low incidence of second brain tumors should therefore not preclude the use of radiotherapy as an effective treatment modality in patients with otherwise uncontrolled pituitary tumors. However, caution is warranted for using RT in benign tumors in childhood and in tumor-prone conditions such as neurofibromatosis [74].

Medical Treatment

The most widely used treatment for pituitary tumors is represented by DA and SSA. DA have been used successfully for PRL-secreting tumors with an overall success rate for approximately 80% in macroprolactinomas [75]. Recent data have demonstrated the expression and function of dopamine receptors not only in endocrine organs but also in endocrine tumors, mainly those belonging to the hypothalamus-pituitary-adrenal axis, and also in the so-called 'neuroendocrine' tumors [76]. Patients with malignant prolactinomas can either present with DA resistance from the beginning or may later on develop resistance to dopamine agonists [75]. DA treatment in patients with PRL-secreting carcinomas have revealed lack of response or tumor progression after treatment [75] and only one patient with partial biochemical response was reported [77]. DA have been used in GH-, ACTH- and TSH-secreting carcinomas, with minimal benefit [6].

SSA, octreotide and lanreotide, that bind with high affinity pituitary tumors expressing the subtype 2 and 5 of somatostatin receptors, have been used to control the syndromes of excess GH, ACTH and TSH with different results but failed to control tumor growth [6]. In patients with malignant GH-secreting tumors, the GH receptor antagonist, pegvisomant, may be helpful in normalizing IGF-1 levels, but no clinical experience is still reported. The role of the recently developed somatostatin analog SOM230 (pasireotide) that exhibit a higher affinity for the somatostatin receptors 1–3 and 5 could have a therapeutic impact that should be evaluated.

Biotherapy and Chemotherapy

The antitumoral action of interferon-α (IFN-α), produced by leukocytes, includes direct effects on proliferation, apoptosis, differentiation, and angiogenesis. In cultured pituitary adenoma cells IFN-α potently suppressed hormone secretion, with an additive effect on octreotide and bromocriptine [78]. In the few patients receiving IFN-α after surgery and RT, however, no significant improvement was noticed [79, 80].

Different chemotherapy protocols have been tested in pituitary carcinomas, generally with no significant effect on tumor size or secretion, and only transient improvement or stabilization in a minority of patients [6]. Vaughan et al. [81] treated a patient with an invasive ACTH-secreting tumor, who had already received maximal RT, with a chemotherapy regimen based on procarbazine, etoposide, and lomustine and achieved stable disease after 1 year. Kasperlik-Zaluska et al. [82] tested a combination of lomustine and doxorubicin in a patient with an aggressive GH-secreting tumor with clinical, hormonal, and radiological objective response that remained stable after 2 years. Reports concerning responses to chemotherapy in patients with pituitary carcinoma are more conflicting. Kaizer et al. [83] treated a patient with ACTH-secreting carcinoma with a combination of cyclophosphamide, doxorubicin, and 5-fluorouracil obtaining initial stabilization and delayed regression of metastatic deposits. Conversely, initial response of a PRL-secreting carcinoma with CNS metastases to lomustine, procarbazine, and etoposide [84] and that of a TSH-secreting tumor to 5-fluorouracil and adriamycin [85] were followed by recurrence and disease progression. Kaltsas et al. [80] reported the results of chemotherapy in 7 patients, 3 with the diagnosis of highly aggressive pituitary tumors and 4 with pituitary carcinomas. The combination of lomustine and 5-fluorouracil, although well tolerated, was followed by a poor response rate overall in terms of tumor shrinkage, with transient clinical responses in some patients. The administration of carboplatin, either alone or in combination with 5-fluorouracil or IFN-α, failed to achieve any response [80]. Patients with existing systemic metastases died within few months after chemotherapy treatment beginning, whereas 2 patients receiving chemotherapy before metastases became evident survived 52 and 63 months, respectively [80].

In recent years, temozolamide, an alkylating compound that depletes 0–6-methyl-guanine-DNA methyltransferase (MGMT) a DNA repair enzyme, has been reported to be an encouraging drug for the treatment of pituitary carcinomas [13, 86–88]. A potential advantage of alkylating drugs, such as temozolamide, relies on the fact that they are not cell-cycle specific and can inhibit all stages of tumor-cell growth;

therefore, patients with slow-growing tumors, such as pituitary tumors, might be better suited to this drug type.

The potential use of temozolamide in pituitary carcinomas needs, however, to be validated by multicenter randomized controlled trials.

Conclusion

Pituitary carcinomas are currently defined as pituitary tumors with cranial or systemic metastases. Most cases first presented as pituitary adenomas which responded to treatment but then escaped to therapy after months to many years. Other cases, conversely, presented since the beginning as aggressive tumors which do not respond to conventional therapy. Several molecular pathogenetic hypotheses have been proposed. It may be nevertheless considered aggressive a pituitary adenomas when there is a change in the behavior of a previous adenoma. Some authors [4, 19] suggest that primary tumors with mitotic activity, as with increased (>3%) MIB-1 labeling index, and/or p53 immunoreactivity, should be defined as 'aggressive adenomas' to denote their aggressive potential and indicating a possibility for future malignant transformation [6].

The current multimodal treatments for pituitary carcinomas include repeated debulking operations to decompress vital structures to provide immediate symptomatic relief, conventional RT or radio-surgery, to achieve temporary local control, and medical treatment, including biotherapy and systemic chemotherapy. However, treatment with all these approaches have been shown to be associated with poor results.

References

1 Kaltsas GA, Grossman AB: Malignant pituitary tumors. Pituitary 1998;1:69–81.
2 Scheithauer BW, Gaffey TA, Lloyd RV, Sebo TJ, Kovacs KT, Horvath E, Yapicier O, Young WF Jr, Meyer FB, Kuroki T, Riehle DL, Laws EL: Pathobiology of pituitary adenomas and carcinomas. Neurosurgery 2006;59:341–353.
3 Selman WR, Laws ER, Scheithauer BW, Carpenter SM: The occurrence of dural invasion in pituitary adenomas. J Neurosurg 1986;64:402–407.
4 Pernicone PJ, Scheithauer BW, Sebo TJ, Kovacs KT, Horvath E, Young WF Jr, Lloyd RV, Davis DH, Guthrie BL, Schoene WC: Pituitary carcinoma: A clinicopathologic study of 15 cases. Cancer 1997;79:804–812.
5 Ragel BT, Couldwell WT: Pituitary carcinoma: a review of the literature. Neurosurg Focus 2004;16:1–9.
6 Kaltsas GA, Panagiotis N, Kontogeorgos G, Buchfelder M, Grossman AB: Diagnosis and management of pituitary carcinomas. J Clin Endocrinol Metab 2005;90:3089–3099.
7 Scheithauer BW, Kurtkaya-Yapicier O, Kovacs KT, Young WF Jr, Lloyd RV: Pituitary carcinoma: a clinicopathological review. Neurosurgery 2005;56:1066–1074.
8 Lopes MB, Scheithauer BW, Schiff D: Pituitary carcinoma: diagnosis and treatment. Endocrine 2005;28:115–121.
9 Gurlek A, Karavitaki N, Ansorge O, Wass JA: What are the markers of aggressiveness in prolactinomas? Changes in cell biology, extracellular matrix components, angiogenesis and genetics. Eur J Endocrinol 2007;156:143–153.

10 Scheithauer BW, Kovacs KT, Laws ER, Randall RV: Pathology of invasive pituitary tumors with special reference to functional classification. J Neurosurg 1986;65:733–744.

11 Roncaroli F, Scheithauer BW, Young WF, Horvath E, Kovacs K, Kros JM, Sarraj SA, Lloyd RV, Faustini-Fustini M: Silent corticotroph carcinoma of the adenohypophysis: a report of five cases. Am J Surg Pathol 2003;27:477–486.

12 Sivan M, Nandi D, Cudlip S: Intramedullary spinal metastases (ISCM) from pituitary carcinoma. J Neurooncol 2006;80:19–20.

13 Fadul CE, Kominsky AL, Meyer LP, Kingman LS, Kinlaw WB, Rhodes CH, Eskey CJ, Simmons NE: Long-term response of pituitary carcinoma to temozolomide: report of two cases. J Neurosurg 2006;105:621–626.

14 Brown RL, Muzzafar T, Wollman R, Weiss RE: A pituitary carcinoma secreting TSH and prolactin: a non-secreting adenoma gone awry. Eur J Endocrinol 2006;154:639–643.

15 Zhu Y, Shahinian H, Hakimian B, Bonert V, Lim S, Heaney A: Temodar: novel treatment for pituitary carcinoma. US Endocr Soc 2004;138:43–45.

16 Kumar K, Wilson JR, Li Q, Phillipson R: Pituitary carcinoma with subependymal spread. Can J Neurol Sci 2006;33:329–332.

17 Tena-Suck ML, Salinas-Lara C, Sanchez-Garcia A, Rembao-Bojorquez D, Ortiz-Plata A: Late development of intraventricular papillary pituitary carcinoma after irradiation of prolactinoma. Surg Neurol 2006;66:527–533.

18 Yamashita H, Nakagawa K, Tago M, Nakamura N, Shiraishi K, Yamauchi N, Ohtomo K: Pathological changes after radiotherapy for primary pituitary carcinoma: a case report. J Neuro-Oncol 2005;75:209–214.

19 Crusius PS, Forcelini CM, Mallmann AB, Silveira DA, Lersch E, Seibert CA, Crusius MU, Carazzo CA, Crusius CU, Goellner E: Metastatic prolactinoma: case report with immunohistochemical assessment for p53 and ki-67 antigens. Arq Neuropsiquiatr 2005;63:864–869.

20 Ceyhan K, Yagmurlu B, Dogan BE, Erdogan N, Bulut S, Erekul S: Cytopathologic features of pituitary carcinoma with cervical vertebral bone metastases: a case report. Acta Cytol 2006;50:225–230.

21 Minniti G, Traish D, Ashley S, Gonsalves A, Brada M: Risk of second brain tumor after conservative surgery and radiotherapy for pituitary adenoma: update after an additional 10 years. J Clin Endocrinol Metab 2005;90:800–804.

22 Brada M, Ford D, Ashley S, Bliss JM, Crowley S, Mason M, Rajan B, Traish D: Risk of second brain tumor after conservative surgery and radiotherapy for pituitary adenoma. Br Med J 1992;304:1343–1346.

23 Tsang RW, Brierley JD, Panzarella T, Gospodarowicz MK, Sutcliffe SB, Simpson WJ: Radiation therapy for pituitary adenoma: treatment outcome and prognostic factors. Int J Radiat Oncol Biol Phys 1994;30:557–565.

24 Erfurth EM, Bulow B, Mikoczy Z, Hagmar L: Incidence of a second tumor in hypopituitary patients operated for pituitary tumors. J Clin Endocrinol Metab 2001;86:659–662.

25 Bliss P, Kerr GR, Gregor A: Incidence of second brain tumours after pituitary irradiation in Edinburgh 1962–1990. Clin Oncol R Coll Radiol 1994;6:361–363.

26 Taylor WA, Uttley D, Wilkins PR: Multiple dural metastases from a pituitary adenoma. Case report. J Neurosurg 1994;81:624–626.

27 Lloyd RV, Kovacs K, Young WF Jr, Farrel WE, Asa SL, Trouillas J, Kontogeorgos G, Sano T, Scheithauer BW, Horvath E, De Lellis RA, Heitz PU: Pituitary tumors; in DeLellis R, Lloyd RV, Heitz PV, Eng C (eds): Introduction. WHO Classification of Tumors of the Endocrine Organs: Pathology and Genetics of Endocrine Organs. Lyon, IARC Press, 2004, pp 10–13.

28 Wilson DF: Pituitary carcinoma occurring as middle ear tumor. Otolaryngol Head Neck Surg 1982;90:665–666.

29 Clayton RN, Farrell WE: Clonality of pituitary tumors: more complicated than initially envisaged? Brain Pathol 2001;11:313–327.

30 Bates AS, Farrell WE, Bicknell EJ, Mc Nicol AM, Talbot AJ, Broome JC, Perrett CW, Thakker RV, Clayton RN: Allelic deletion in pituitary adenomas reflects aggressive biological activity and has potential value as a prognostic marker. J Clin Endocrinol Metab 1997;82:818–824.

31 Gaffey TA, Scheithauer BW, Lloyd RV, Burger PC, Robbins P, Fereidooni F, Horvath E, Kovacs K, Kuroki T, Young WF Jr, Sebo TJ, Riehle DL, Belzberg AJ: Corticotroph carcinoma of the pituitary: a clinico-pathological study. Report of four cases. J Neurosurg 2002;6:360.

32 Zafar MS, Mellinger RC, Chason JL: Cushing's disease due to pituitary carcinoma. Henry Ford Hosp Med J 1984;32:61–66.

33 Zahedi A, Booth GL, Smyth HS, Farrell WE, Clayton RN, Asa SL, Ezzat S: Distinct clonal composition of primary and metastatic adrenocorticotrophic hormone producing pituitary carcinoma. Clin Endocrinol 2001;55:549–556.

34 Chesnokova V, Zonis S, Kovacs K, Ben-Shlomo A, Wawrowsky K, Bannykh S, Melmed S. p21Cip1 restrains pituitary tumor growth. Proc Natl Acad Sci USA 2008;105:17498–17503.

35 Rickert CH, Scheithauer BW, Paulus W: Chromosomal aberrations in pituitary carcinoma metastases. Acta Neuropathol (Berl) 2001;102:117–120.

36 Suhardja A, Kovacs K, Rutka J: Genetic basis of pituitary adenoma invasiveness: a review. J Neurooncol 2001;52:195–204.

37 Shimon I, Melmed S: Pituitary tumour pathogenesis. J Clin Endocrinol Metab 1997;82:1675–1681.

38 Farrell WE, Clayton RN: Molecular pathogenesis of pituitary tumors. Front Neuroendocrinol 2000;21:174–198.

39 Bates AS, Farrell WE, Bicknell EJ, McNicol AM, Talbot AJ, Broome JC, Perrett CW, Thakker RV, Clayton RN: Allelic deletion in pituitary adenomas reflects aggressive biological activity and has potential value as a prognostic marker. J Clin Endocrinol Metab 1997;82:818–824.

40 Pei L, Melmed S, Scheithauer B, Kovacs K, Prager D: H-ras mutations in human pituitary carcinoma metastases. J Clin Endocrinol Metab 1994;78:842–846.

41 Cryns VL, Alexander JM, Klibanski A, Arnold A: The retinoblastoma gene in human pituitary tumors. J Clin Endocrinol Metab 1993;77:644–646.

42 Takino H, Herman V, Weiss M, Melmed S: Purine-binding factor (nm23) gene expression in pituitary tumors: marker of adenoma invasiveness. J Clin Endocrinol Metab 1995;80:1733–1738.

43 Gandour-Edwards R, Kapadia SB, Janecka IP, Martinez AJ, Barnes L: Biologic markers of invasive pituitary adenomas involving the sphenoid sinus. Modern Pathology 1995;8:160–164.

44 Jin L, Qian X, Kulig E, Sanno N, Scheithauer BW, Kovacs K, Young WF, Lloyd RV: Transforming growth factor-ß, transforming growth factor-ß receptor II, and p27Kip1 expression in nontumorous and neoplastic human pituitaries. Am J Pathol 1997;151:509–519.

45 Lidhar K, Korbonits M, Jordan S, Khalimova Z, Kaltsas G, Lu X, Clayton RN, Jenkins PJ, Monson JP, Besser GM, Lowe DG, Grossman AB: Low expression of the cell cycle inhibitor p27Kip1 in normal corticotroph cells, corticotroph tumors, and malignant pituitary tumors. J Clin Endocrinol Metab 1999;84:3823–3830.

46 Meij BP, Lopes MB, Ellegala DB, Alden TD, Laws ER: The long-term significance of microscopic dural invasion in 354 patients with pituitary adenomas treated with transsphenoidal surgery. J Neurosurg 2002;96:195–208.

47 Turner HE, Wass JA: Are markers of proliferation valuable in the histological assessment of pituitary tumors? Pituitary 1999;1:147–151.

48 Thapar K, Kovacs K, Muller P: Clinical-pathologic correlations of pituitary tumors. Baillieres Clin Endocrinol Metab 1995;9:243–270.

49 Vidal A, Kovacs K, Howarth E, Scheithauer BW, Kuroki T, Lloyd RV: Microvessel density in pituitary adenomas and carcinomas. Virchows Arch 2001;438:595–602.

50 Zuber M, Tan EM, Ryoji M: Involvement of proliferating cell nuclear antigen (cyclin) in DNA replication in living cells. Mol Cell Biol 1989;9:57–66.

51 Tziortzioti V, Ruebel KH, Kuoki T, Jin L, Scheithauer BW, Lloyd RV: Analysis of ß-catenin mutations and α-, ß- and γ-catenin expression in normal and neoplastic human pituitary tissues. Endocr Pathol 2001;12:125–136.

52 Ezzat S, Horvath E, Kovacs K, Smyth HS, Singer W, Asa SL: Basic fibroblast growth factor expression by two prolactin and thyrotropin producing pituitary adenomas. Endocr Pathol 1995;6:125–134.

53 Abbass SA, Asa SL, Ezzat S: Altered expression of fibroblast growth factor in human pituitary adenomas. J Clin Endocrinol Metab 1997;82:1160–1166.

54 Onguru O, Scheithauer BW, Kovacs K, Vidal S, Jin L, Zhang S, Ruebel KH, Lloyd RV: Analysis of epidermal growth factor receptor and activated epidermal growth factor receptor expression in pituitary adenomas and carcinomas. Modern Pathol 2004;17:772–780.

55 Roncaroli F, Nose V, Scheithauer BW, Kovacs K, Horvath E, Young WF Jr., Lloyd RV, Bishop MC, Hsi B, Fletcher JA: Gonadotropic pituitary carcinoma: HER-2/neu expression and gene amplification: report of two cases. J Neurosurg 2003;99:402–408.

56 Nose V, Scheithauer BW, Lloyd RV, Kovacs K, Horvath E, Kroll T, Fletcher J: Her-2/neu genomic aberrations in pituitary carcinoma. Endocr Pathol 2001;12:231–232.

57 Chae YS, Flotte T, Hsu DW, Preffer F, Hedley-Whyte ET: Flow cytometric DNA ploidy and cells phase fractions in recurrent human pituitary adenomas: a correlative study of flow cytometric analysis and the expression of proliferating cell nuclear antigen. Gen Diagn Pathol 1996;142:89–95.

58 Gaffey TA, Scheithauer BW, Leech RW, Blick K, Kovacs K, Horvath E, Weaver AL, Lloyd RV, Ebersold M, Laws ER, De Bault LE: Pituitary adenoma: a DNA flow cytometric study of 192 clinicopathologically characterized tumors. Clin Neuropathol 2005;24:56–63.

59 Riss D, Jin L, Qian X, Bayliss J, Scheithauer BW, Young WF Jr, Vidal S, Kovacs K, Raz A, Lloyd RV: Differential expression of galectin-3 in pituitary tumors. Cancer Res 2003;63:2251–2255.

60 Vidal A, Kovacs K, Bell D, Horvath E, Scheithauer BW, Lloyd RV: Cyclooxygenase-2 expression in human tumors. Cancer 2003;97:2814–2821.

61 Onguru O, Scheithauer BW, Kovacs K, Vidal S, Jin L, Zhang S, Ruebel K, Lloyd RV: Analysis of COX-2 and thromboxane synthase expression in pituitary adenomas and carcinomas. Endocr Pathol 2004;15:17–27.

62 Kaltsas GA, Mukherjee JJ, Grossman AB: The value of radiolabelled MIBG and octreotide in the diagnosis and management of neuroendocrine tumors. Ann Oncol 2001;12:47–50.

63 Greenman Y, Woolf P, Coniglio J, O'Mara R, Pei L, Said JW, Melmed S: Remission of acromegaly caused by pituitary carcinoma after surgical excision of growth hormone-secreting metastases detected by 111-indium pentetreotide scan. J Clin Endocrinol Metab 1996;81:28–1633.

64 Hurel S, Harris PE, Greenman Y, Woolf P, Coniglio J: Remission of acromegaly caused by pituitary carcinoma after surgical excision of growth hormone-secreting metastasis detected by 111-indium pentetreotide scan. J Clin Endocrinol Metab 1996; 81:1628–1633.

65 Garrao AF, Sobrinho LG, Oliveira P, Bugalho MJ, Boavida JM, Raposo JF, Loureiro M, Limbert E, Costa I, Antunes JL: ACTH-producing carcinoma of the pituitary with haematogenic metastases. Eur J Endocrinol 1997;137:176–180.

66 Dayan C, Guilding T, Hearing S, Thomas P, Nelson R, Moss T, Bradshaw J, Levy A, Lightman S: Biochemical cure of recurrent acromegaly by resection of cervical spinal canal metastases. Clin Endocrinol (Oxf) 1996;44:597–602.

67 Eriksson B, Bergstrom M, Sundin A, Juhlin C, Orlefors H, Oberg K, Langstrom B: The role of PET in localization of neuroendocrine and adrenocortical tumors. Ann NY Acad Sci 2002;970:159–169.

68 Vaquero J, Herrero J, Cincu R: Late development of frontal prolactinoma after resection of pituitary tumor. J Neurooncol 2003;64:255–258.

69 Martin NA, Hales M, Wilson CB: Cerebellar metastases from a prolactinoma during treatment with bromocriptine. J Neurosurg 1981;55:615–619.

70 Epstein A, Epstein BS, Molho L, Zimmerman HM: Carcinoma of the pituitary gland with metastases to the spinal cord and roots of the cauda equina. J Neurosurg 1964;21:846–853.

71 Swords FM, Allan CA, Plowman PN, Sibtain A, Evanson J, Chew SL, Grossman AB, Besser GM, Monson JP: Stereotactic radiosurgery. XVI. A treatment for previously irradiated pituitary adenomas. J Clin Endocrinol Metab 2003;88:5334–5340.

72 Winkelmann J, Pagotto U, Theodoropoulou M, Tatsch K, Saeger W, Muller A, Arzberger T, Schaaf L, Schumann EM, Trenkwalder C, Stalla GK: Retention of dopamine 2 receptor mRNA and absence of the protein in craniospinal and extracranial metastases of a malignant prolactinoma: a case report. Eur J Endocrinol 2002;146:81–88.

73 Kaltsas G, Rockall A, Papadogias D, Reznek R, Grossman AB: Recent advances in radiological and radionuclide imaging and therapy of neuroendocrine tumors. Eur J Endocrinol 2004;151:15–27.

74 Evans DGR, Birch JM, Ramsden RT, Sharif S, Baser ME: Malignant transformation and new primary tumors after therapeutic radiation for benign disease: substantial risks in certain tumor prone syndromes. J Med Genet 2006;43:289–294.

75 Gillam MP, Molitch ME, Lombardi G, Colao A: Advances in the treatment of prolactinomas. Endocr Rev 2006;27:485–534.

76 Pivonello R, Ferone D, Lombardi G, Colao A, Lamberts SW, Hofland LJ: Novel insights in dopamine receptor physiology. Eur J Endocrinol 2007; 156(suppl):S13–S21.

77 Popovic EA, Vattuone JR, Siu KH, Busmanis I, Pullar MJ, Dowling J: Malignant prolactinomas. Neurosurgery 1991;29:127–130.

78 Hofland LJ, de Herder WW, Waaijers M, Zuijderwijk J, Uitterlinden P, van Koetsveld PM, Lamberts SW: Interferon-alfa-2a is a potent inhibitor of hormone secretion by cultured human pituitary adenomas. J Clin Endocrinol Metab 1999;84:3336–3343.

79 Ahmed M, Kanaan I, Alarifi A, Ba-Essa E, Saleem M, Tulbah A, McArthur P, Hessler R: ACTH-producing pituitary cancer: experience at the King Faisal Specialist Hospital, Research Centre. Pituitary 2000;3:105–112.

80 Kaltsas GA, Mukherjee JJ, Plowman PN, Monson JP, Grossman AB, Besser GM: The role of cytotoxic chemotherapy in the management of aggressive and malignant pituitary tumors. J Clin Endocrinol Metab 1998;83:4233–4238.

81 Vaughan NJA, Laroche CM, Goodman I, Davies MJ, Jenkins JS: Pituitary Cushing's disease arising from a previously non-functional corticotrophic chromophobe adenoma. Clin Endocrinol (Oxf) 1985;22:147–153.

82 Kasperlik-Zaluska AA, Wislawski J, Kaniewska J, Zborzil J, Frankiewicz E, Zgliczynski S: Cytostatics for acromegaly: marked improvement in a patient with an invasive pituitary tumour. Acta Endocrinol (Copenh) 1987;116:347–349.

83 Kaizer FE, Orth DN, Mukai K, Oppenhaimer JH: A pituitary parasellar tumour with extracranial metastases and high, partially suppressible levels of adrenocorticotropin and related peptides. J Clin Endocrinol Metab 1983;57:649–653.

84 Peterson T, McFarlane IA, McKenzie JM, Shaw MD: Prolactin secreting pituitary carcinoma. J Neurol Neurosurg Psychol 1992;55:1205–1206.

85 Mixson AJ, Friedman TC, Katz DA, Feuerstein IM, Taubenberger JK, Colandrea JM, Doppman JL, Oldfield EH, Weintraub BD: Thyrotropin-secreting pituitary carcinoma. J Clin Endocrinol Metab 1993;76:529–533.

86 Kovacs K, Horvath E, Syro LV, Uribe H, Penagos LC, Ortiz LD, Fadul CE: Temozolomide therapy in a man with an aggressive prolactin-secreting pituitary neoplasm: morphological findings. Human Pathology 2007;38:185–189.

87 Syro LV, Uribe H, Penagos LC, Ortiz LD, Fadul CE, Horvath E, Kovacs K: Antitumor effects of temozolomide in a man with a large, invasive prolactin-producing pituitary neoplasm. Clin Endocrinol 2006;65:549–553.

88 Lim S, Shahinian H, Maya MMM, Yong W, Heaney AP: Temozolomide: a novel treatment for pituitary carcinoma. Lancet Oncol 2006;7:518–520.

Annamaria Colao, MD, PhD
Department of Molecular and Clinical Endocrinology and Oncology, 'Federico II' University
via S. Pansini 5
IT–80131 Napoli (Italy)
Tel. +39 0 81 7462132, Fax +39 081 5465443, E-Mail colao@unina.it

Modern Imaging of Pituitary Adenomas

Michael Buchfelder · Sven-Martin Schlaffer

Department of Neurosurgery, University of Erlangen-Nürnberg, Erlangen, Germany

Abstract

Before CT and MR imaging were available, pituitary mass lesions could not directly be depicted. Conventional X-ray studies, pneumoencephalography and catheter angiography could only reveal indirect sings of a space-occupying intra-, para- or suprasellar lesion. Only when CT scanning and later on low- and high-field magnetic resonance imaging were introduced into clinical practice, were the radiological diagnoses of pituitary adenomas refined, a much better morphological contribution to differential diagnosis of a lesion obtained, and the effects of therapies, such as medical treatments, irradiation and surgery, could be monitored. To date, magnetic resonance tomography is generally considered the imaging method of choice for its premium soft tissue contrast. The goal of all imaging studies is to indicate precisely the localization and nature of the sella region tumor, its extension in relation to the various surrounding structures, its structure and its enhancement in order to help in the differential diagnosis and treatment planning.

Copyright © 2010 S. Karger AG, Basel

Magnetic Resonance Imaging

Historically, MRI techniques for detection and depiction of pituitary adenomas have witnessed rapid evolution ranging from the onset of non-contrast MRI studies in late 1980s to introduction of thin section contrast MRI scans in early 1990s. With the wide availability of MR scanners, this technique is to date and for the near future the standard imaging procedure for sellar processes since it allows evaluation of the sella and perisellar lesions with high soft tissue contrast and excellent anatomical resolution, avoids ionizing radiation and has multiplanar capabilities [3, 12]. Usually, the contrast medium is injected intravenously as a bolus and the postcontrast MR scans are thus obtained with some delay.

Standard MR imaging of the sella and perisellar region should consist of 2–3 mm thin section T_1-weighted TSE images with and without contrast enhancing medium (i.e. gadolinium) in sagittal and coronal planes on a high (512) matrix. T_2-weighted TSE or FSE images can provide additional information about the cystic components of the lesion, regressive changes and hemorrhage. Contrast medium differentiates between the pituitary gland, blood vessels and other tissues [3].

If clinical features suggest the presence of a functioning pituitary adenoma and regular standardized MR imaging is not able to detect a pituitary adenoma, dynamic MR sequences might be added. These scans are taken in 20- to 30-second intervals after intravenous injection of gadolinium contrast agents. The pituitary stalk and posterior lobe enhance approximately 20 s after injection and the anterior pituitary after 80 s [3, 12].

In addition, a MR angiography can provide detailed information about the arterial blood vessels surrounding the sella region, such as potential displacements and encasements.

The interpretation of MR imaging requires a profound knowledge of the normal anatomy of the sellar region and of the normal anatomical variants, such as shape of the bony sella, structure of the sellar diaphragm, pneumatization of the sphenoid sinus and venous drainage at the sellar floor [12, 19]. The normal anterior pituitary usually has a isointense signal compared with the white matter of the temporal lobe on T_1-weighted images (fig. 1), whereas the posterior lobe has a hyperintense signal ('bright spot'). Furthermore, there are several potential artifacts that need to be considered, such as pixel-to-pixel inhomogeneity, partial volume effects, chemical shifting, and others. Moreover, the reader should be familiar with the typical appearances and major variants of tumors, hyperplasias, malformations, vascular lesions and inflammations, which occur in the sellar and perisellar region [3, 12].

In the perisellar region, MR imaging can provide information about the size, localization and extent of a lesion, may it be para- (lateral), supra- (superior), retro- (posterior to the dorsum sellae) or infrasellar (sphenoidal). Moreover, one can suspect potential invasion of a pituitary adenoma into either the sphenoid cavity or cavernous sinus.

Imaging of Pituitary Adenomas

Pituitary microadenomas are defined as lesions with a diameter below 10 mm and usually are localized within sella. They can be reliably visualized directly once the tumor diameter exceeds 4 mm, mostly as distinct rounded, oval, triangular or flattened hypointense regions, hypointense to the anterior gland, but isointense to the grey matter of the temporal cortex on T_1-weighted images (fig. 2). If the microadenoma is isointense to the gland on T_1-weighted images, other sequences may be required for direct depiction. Indirect signs, such as deviation of the infundibulum, upward bulging of the sellar content into the optico-chiasmatic cistern or a localized depression of the sellar floor may be indicating a small space-occupying sellar lesion. Hemorrhagic or cystic changes within an adenoma may alter its signal intensity so that a minority is hyperintense in the T_1-weighted images. The appearance of

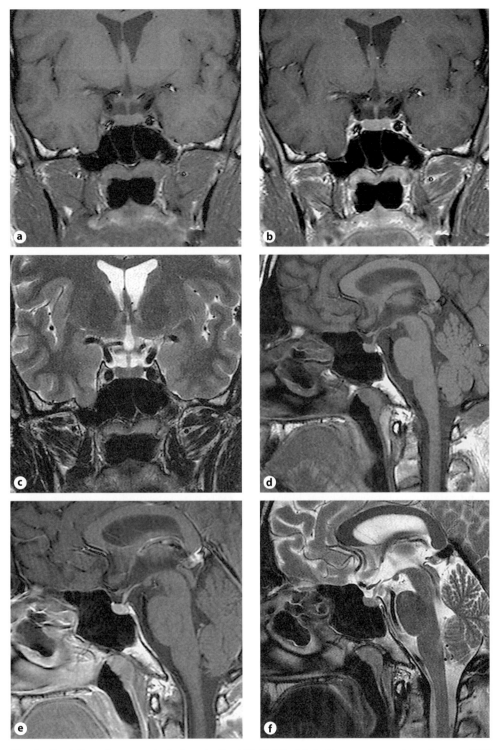

Fig. 1. Depiction of a normal pituitary in the MRI (1.5 Tesla). In the upper panel (**a–c**) coronal MR images of the normal pituitary gland are shown in nonenhanced T_1-weighted images (**a**), T_1-weighted images with contrast medium (**b**) and T_2-weighted images (**c**). In the lower panel, the respective sagittal planes of the same patient are displayed using identical imaging modalities (**d–f**).

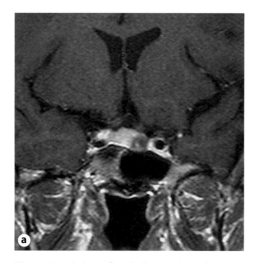 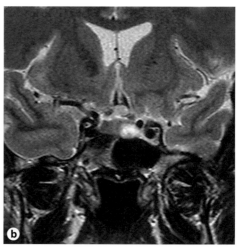

Fig. 2. Depiction of a pituitary microadenoma (1.5 Tesla). In the coronal T_1-weighted images with contrast enhancement (**a**), the microadenoma appears as a distinct zone of lower signal intensity. In the T_2–weighted TSE images (**b**), the pituitary microadenoma is visible as a hyperintense lesion within the sella.

microadenomas on T_2-weighted FSE or TSE images is more variable. If no adenoma is visible on non-enhanced scans, the use of paramagnetic contrast medium can be helpful. Microadenomas usually are depicted as areas of relative hypo-enhancement. The dose of gadolium has to be selected very carefully. Dynamic scanning and the use of delayed post-contrast images can depict some microadenomas, which have not been visible with routine parameter images. There is a huge potential of mis-interpretation, since such apparent tiny regions of altered signal intensity within the pituitary gland might also represent other lesions, e.g. cysts pars intermedia or may be due to artifacts, such as partial volume effects. Thus, the findings have to be appreciated with considerable care [3, 12].

Pituitary macroadenomas are defined as sellar mass lesions with a diameter greater than 10 mm. They can be reliably visualized with routine MR imaging and, thus, the task is rather to describe precisely the extent, localization, structure and relation to the surrounding anatomical structures that helps with the differential diagnosis. It is characteristic for pituitary adenomas the bulk of the lesion is within the sella. The tumors may be inhomogeneous, since regressive changes may be present (fig. 3). Mostly the extension is superiorly, into the optico-chiasmatic cistern and may reach up to the Foramen of Monro, producing hydrocephalus. There may be extension downwards, into the sphenoid sinus with extension of the sella or even perforation of the sellar floor which is indicative of invasion. Lateral tumor expansion is also common and frequently the question arises if the tumor is still encases or invasive [1]. Only total encasement of the carotid artery, with tumor clearly visible lateral of the vessel proves invasion. Even grotesque lateral extension sometimes turns out not to

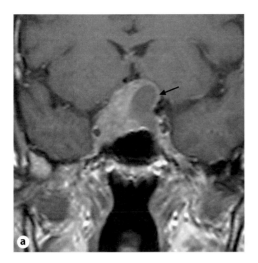 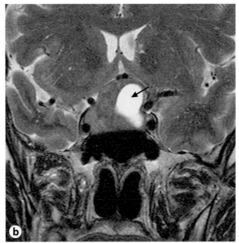

Fig. 3. Depiction of a pituitary macroadenoma (1.5 Tesla). In the native coronal T_1-weighted images (**a**), the macroadenoma appears as a space-occupying intra- and suprasellar tumor that has a mixed signal intensity. In the T_2-weighted TSE images (**b**), the cystic portions of the macroadenoma show a bright signal (arrows), indicating protein-rich content.

have resulted in invasion. In several instances, imaging with very high field strength (3 Tesla) offers direct depiction of the invasion site which was not shown by 1.5 Tesla images [17, 18]. Normally, a compression, flattening and distortion of the normal pituitary gland can be seen, preferably to the posterior or lateral periphery of the tumor. This is particularly well shown on the contrast enhanced images, on which the pituitary enhances quicker and better than the adenoma. Pituitary adenomas are virtually never invasive into the cavernous sinus that is covered by a thin layer of compressed by the displaced pituitary and have their major parasellar extension on the contralateral side [3]. Pituitary surgeons regularly appreciate this growth pattern. The amount of compression and deformation of the optic chiasma is best appreciated on the coronal sections, although with grotesque suprasellar extension, the visual pathways may escape direct depiction.

Postoperative Magnetic Resonance Imaging

A baseline postoperative imaging investigation should be performed some 2–3 months after surgery, preferably with standard sequences, comparable to the initial, pre-treatment MR scan, unless the clinical situation requires earlier postoperative imaging. After total trans-sphenoidal resection of a large pituitary adenoma, the tumor is no longer visible. The space previously occupied by the tumor is filled with CSF. The decompressed optic chiasm and the stretched infundibulum become

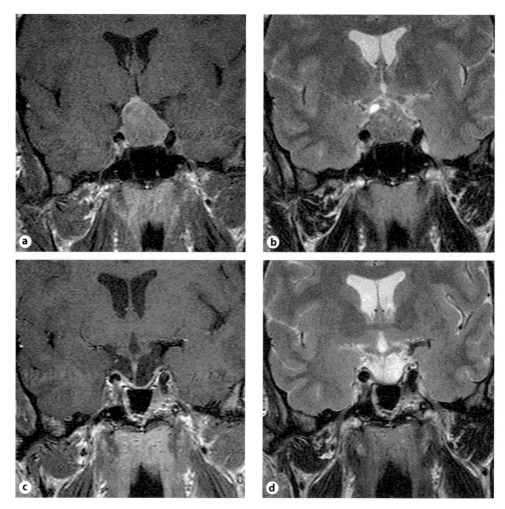

Fig. 4. Pre- and postoperative images of a pituitary macroadenoma (1.5 Tesla): In the coronal T_1-weighted (**a**) and in the T_2-weighted (**b**) images, the macroadenoma appears as a homogeneous space-occupying intra- and suprasellar lesion. Three months following trans-sphenoidal surgery, delayed postoperative imaging reveals complete tumor resection in both T_1 (**c**) and T_2-weighted (**d**) images with chiasm displayed and the pituitary stalk reaching flattened remnants of the normal gland at the bottom of the sella. The normal postoperative situation is also called 'empty sella' or cisternal herniation.

more clearly visible. The typical postoperative appearance after such a total resection is a more or less expressed cisternal herniation or 'empty sella' (fig. 4). In patients with asymmetrical, large tumors, particularly those who harbor invasive tumors, comparison of pre- and postoperative images reveal residual tumor that could not be resected (fig. 5), harboring the same signal intensity as the tumor preoperatively [3]. Some caution has to be applied: The postoperative cavity is often filled with blood a few hours and days after surgery and the resorption of

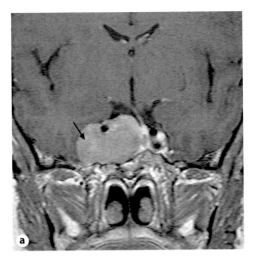 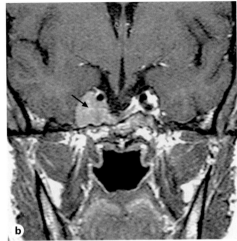

Fig. 5. Partial resection of an invasive pituitary adenoma (1.5 Tesla). A large, homogeneous, intra-, supra- and parasellar pituitary adenoma is visible preoperatively in the T_1-weighted coronal MR sections (**a**) with tumor also laterally from the carotid artery (arrows), indicating invasion. Three months postoperatively (**b**), the intra- and suprasellar tumor portions are resected, resulting in cisternal herniation. However, there is residual tumor within the cavernous sinus. The hyperintense, deformed normal pituitary is localized on the right side and is still visible.

this material takes several weeks [6]. Thus, in an early scan, a considerable space-occupying lesion might still be visible, which appears quite threatening. There are a variety of postoperative artifacts and structural changes which need to be differentiated from residual pituitary tumor. Drilling of the skull base, which is frequently required during such operations, may leave metal debris, which produces particularly nasty artifacts. Autologous or heterologous implants, such as fascia lata, fat and muscle, may have been used to reconstruct the sellar floor or even been introduced into the tumor cavity to prevent a CSF leak. Fluid retentions and mucosal regenerates within the paranasal cavities are other types of structures that should not be confused with residual or recurrent tumor [4, 9, 11, 14, 15]. In the further follow-up, it is essential that the whole series of postoperative scans are obtained using identical scanning parameters. The entire set needs to be evaluated rather than only the current and previous examination, since most pituitary adenomas are slowly growing and changes become more apparent when time advances. The baseline delayed postoperative image, obtained 2–3 months after the surgery plays a pivotal role. Sometimes, it is initially impossible to differentiate implanted materials and scarring from residual tumor. Only progression of tumor over time, while implants regress, resolves the diagnostic dilemma. The same principles apply to the follow-up of pituitary tumors after irradiation or during medical treatments. Considering follow-up intervals, a safe practice would be to obtain a repeat MR scan every year following the initial delayed postoperative baseline investigation.

In patients with residual tumor this practice should be continued, while in patients in whom no tumor is visible after 5 years, the follow-up intervals could be stretched to longer periods [8,16]. Volumetry requires a specific software and is extremely time consuming [2]. Thus, to date, it is still reserved for research projects and not used routinely in clinical practice.

Intraoperative MR Imaging

While standardized pre- and postoperative imaging is widely available, only a few surgical centers offer intraoperative MR imaging at present. The major asset is that in pituitary surgery intraoperative imaging can provide information about residual tumor that is hidden to the surgeon initially, but can be attacked while the patient is still in the operating room. Therefore, the technique helps increase the rate of complete tumor resections. In contrast to delayed, postoperative images, blood and residual tumor need to be differentiated intraoperatively. Thus, in contrast to routine delayed postoperative MR scanning, sagittal and coronal T_2-weighted TSE sequences are the ideal sequences to fulfill this task [7]. The surgeon needs to compare the pre- and intraoperative situation. Especially in large asymmetric, 'giant' adenomas where intraoperative estimation of the resection extent is difficult, tumor residuals can be localized. One expects that a total resection can be achieved in approximately an additional 30% of the patients, but this obviously depends on the patient selected to undergo surgery when this gadget is available [3, 7].

Differential Diagnosis

MR imaging may particularly help with the differential diagnosis of lesions in and around the sella turcica, and a huge body of literature covers this issue [12, 19]. Briefly, craniopharyngiomas may have an intrasellar component, but are usually predominantly localized suprasellarly. They frequently consist of cystic, solid and calcified portions. Only a small minority of craniopharyngiomas has only intrasellar extension. Rathke's cleft cysts lack solid and cystic portions. The final differentiation from cystic craniopharyngioma requires histological investigation of the epithelial membrane. Suprasellar meningiomas have an attachment to the planum sphenoidale. While in pituitary adenomas, the normal pituitary is dislocated into the superolateral periphery of the tumor, in suprasellar meningiomas, the normal gland is usually compressed at the basis of the sella. Parasellar meningiomas distend the cavernous sinus and narrow the carotid artery. Chordomas of the clivus distend the clivus and dorsum sellae (fig. 6). In patients with suprasellar germinomas and in those with infundibulohypophysitis, the infundibulum is enlarged and enhances with contrast medium, while hypothalamic hamartomas are classical

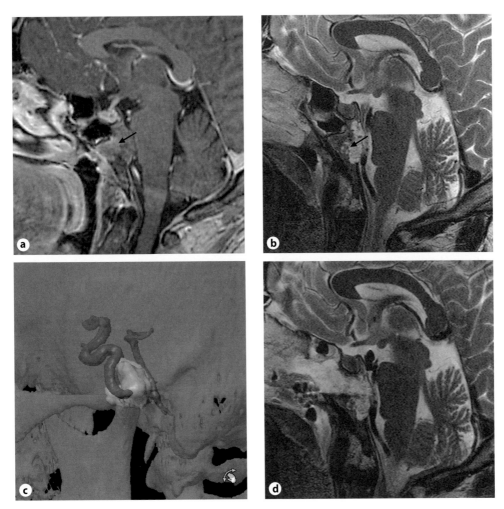

Fig. 6. Imaging of a chordoma of the clivus (1.5 Tesla). A 33-year-old patient presented with a lesion (arrows) of the clivus in front of the basilar artery in the T_1- (**a**) and T_2-weighted (**b**) images. Using multimodal (CT and MRI) reconstructions, a pseudo 3-D image was created (**c**) which also visualized the dataset that was used for neuronavigation. The chordoma as well as the carotid and basilar artery are shown. The tumor was resected via the trans-sphenoidal approach. The intraoperative MR (T_2-weighted TSE sequences) depicts the tumor resection (**d**) and shows some blood within the sphenoid sinus.

prototypes of nonenhancing lesions. There are many small incidental lesions picked up by sophisticated imaging, which lack clinical correlates. More than ever is the physician to date in such situations challenged with the difficult decision masking when to intervene.

However, only the consideration of clinical features and laboratory findings together with the imaging results provides enough information to establish a reliable differential diagnosis [12, 13, 19].

Computerized Tomography

While a 'normal' axial CT scan of the brain that is, for example, used to rule out a space-occupying intracerebral hematoma after a trauma and employs rather thick slices is unable to detect small pituitary tumors or microadenomas, it can reveal large pituitary macroadenoms and other perisellar tumors like meningeomas or craniopharyngeomas. This happens mainly when a CT is obtained for some other reason, e.g. after a head injury, and tumors are detected as an incidental finding. However, a proper thin-collimation CT in the coronal sections can demonstrate microadenomas very well and precisely, especially when contrast medium is applied.

With the availability of the multislice CT scanner (64 or 128 slices) generation scanning became very fast and image quality improved considerably. Especially when contraindications for MRI exist (e.g. claustrophobia or pacemaker), contrast-enhanced CT scans can provide reliable information about sellar region tumors [19] when a three-dimensional dataset is available and sagittal and coronal reconstructions are possible. In contrast to MR studies, the strength of a CT scan lies in the precise information it delivers about osseous structures and paranasal sinuses (e.g. septation of sinuses, calcification of craniopharyngeomas, destruction of skull base by malignomas). Information about the structure surgical approaches. Furthermore, a preoperative three-dimensional dataset can be used for neuronavigation during surgery in particularly cases (e.g. McCune-Albright syndrome).

Unfortunately, computerized tomography is associated with radiation exposure, so that as a standard imaging procedure MR is clearly preferable [12].

SPECT/PET

While this article clearly focuses on the present use of MR imaging, molecular imaging needs to be mentioned in the context of pituitary tumor imaging: The presence of some dopamine receptors can be detected in almost all patients with hormonally active as well as in nonfunctioning pituitary adenomas, only in prolactin- and in some GH-secreting adenomas can high numbers of high-affinity D2 receptors be detected. Different SPECT methods are used: ^{123}I-S-(K)-N-[(1-ethyl-2-pyrrolidinyl)methyl]-2-hydroxy-3-iodo-6-methoxybenzamide (^{123}I-IBZM) and ^{123}I-epidepride. ^{123}I-epidepride is generally superior to ^{123}I-IBZM for the visualization of D2 receptors on pituitary macroadenomas. However, the mere presence of receptors does not prove their functionality and does not provide information about post translational processing. Thus, ^{123}I-IBZM and ^{123}I-epidepride scintigraphy are not suitable investigations to predict the response to dopaminergic treatment in pituitary tumor patients. The techniques might allow discrimination of nonfunctioning pituitary macroadenomas from other nontumor pathologies in the sellar region. Dopamine D2 receptors on pituitary tumors can also be studied using positron emission tomography with

^{11}C-N-raclopride and ^{11}C-N-methylspiperone [10]. For their high costs and poor special resolution, these investigations are to date not used in daily clinical practice, but are undoubtedly valuable tools for research purposes. Earlier hope that they might be helpful in aiding nonsurgical treatment options in pituitary tumors have hitherto been disappointed.

Conclusions

For the investigation of pituitary adenomas to date standard imaging consists of high-field MR scanning with T_1-weighted images and a field strength of 1.5 Tesla in sagittal and coronal planes, together with a T_2-weighted sequence in the axial plane. Pre- and postcontrast images are recommended. With this protocol, the differential diagnosis and appropriate follow-up investigations can be performed. Whether the use of 'dynamic sequences' increases tiny lesions unrecognized by standard imaging, remains a matter of controversy. However, images obtained with a 3-Tesla scanner, can sometimes depict invasion sites of tumors into the cavernous sinus. For follow-up investigations, it is important to compare the whole set of investigations, from pretreatment up to the most recent follow-up images to determine whether a tumor is stable, decreased or progressive in size. The strength of CT is the depiction of calcification and structural bony abnormalities. It is only used in selected cases but can still be very valuable for surgical navigation planning. CT is still the second choice method to directly depict a tumor if, for example, the presence of a cardiac pacemaker prohibits acquisition of MR data.

All other modern imaging techniques like PET or SPECT are still experimental and rarely used for clinical decision-making. Earlier hope that they might be helpful in aiding nonsurgical treatment options in pituitary tumors have hitherto been disappointed.

References

1 Ahmadi J, North CM, Segall HD, Zee CS, Weiss MH: Cavernous sinus invasion by pituitary adenomas. AJR 1986;148:257–262.
2 Benesch H, Felber SR, Finkenstedt G, Kremser C, Stockhammer G, Aichner FT: MR volumetry for monitoring intramuscular bromocriptine treatment in macroprolactinomas. J Comput Assist Tomogr 1995;19:866–870.
3 Bonneville JF, Bonneville F, Cattin F: Magnetic resonance imaging of pituitary adenomas. Eur Radiol 2005;15:543–548.
4 Chakabortty S, Oi S, Yamaguchi M, Tamaki N, Matsumoto S: Growth-hormone producing pituitary adenomas: MR characteristics and pre- and postoperative evaluation. Neurol Med Chir 1993;33:81–85.
5 de Herder WW, Reijs AEM, Feelders RA, van Aken MO, Krenning EP, van der Lely AJ, Kwekkeboom DJ: Diagnostic imaging of dopamine receptors in pituitary adenomas. Eur J Endocrinol 2007;156:S53–S56.
6 Dina TS, Feaster SH, Laws ER, Davis DO: MR of the pituitary gland postsurgery: serial MR studies following transsphenoidal resection. AJNR 1993;14:763–769.

7 Fahlbusch R, Keller von B, Ganslandt O, Kreutzer J, Nimsky C: Transsphenoidal surgery in acromegaly investigated by intraoperative high-field magnetic resonance imaging. Eur J Endocrinol 2005;153:239–248.

8 Greenman Y, Quaknine G, Veshchev I, Reider-Groswasser II, Segev Y, Stern N: Postoperative surveillance of clinically non-functioning pituitary macroadenomas: markers of tumour quiescence and regrowth. Clin Endocrinol 2003;58:763–769.

9 Hald JK, Nahstadt PH, Kollovold T, Bakke SJ, Skalpe IO: MR imaging of pituitary macroadenomas before and after transsphenoidal surgery. Acta Radiol 1992;33:369–399.

10 Khan S, Llloyd C, Szysko T, Win Z, Rubello D, Al-Nahhas A: PET imaging in endocrine tumors. Minerva Endocrinol 2008;33:41–42.

11 Mikhael MA, Ciric IS: MR imaging of pituitary macroadenomas before and after surgical and/or medical treatment. J Comp Assist Tomogr 1988;12:441–445.

12 Naidich MJ, Russell EJ: Current approaches to imaging of the sellar region and pituitary. Endocrinol Metab Clin North Am 1999;28:45–79.

13 Ruscalleda J: Imaging of parasellar lesions. Eur Radiol 2005;15:549–559.

14 Steiner E, Knosp E, Herold CJ, Kramer J, Stiglbauer R, Staniszewski K, Imhof H: Pituitary adenomas: findings of postoperative MR images. Radiology 1992;185:521–527.

15 Steiner E, Math G, Knosp E, Mosbeck G, Kramer J, Herold CJ: MR-appearance of the pituitary gland before and after resection of pituitary macroadenomas. Clin Radiol 1994;49:524–530.

16 Soto-Ares G, Crtet-Rudelli C, Assaker R, Boulinguez A, Dubest C, Dewailly D, Pruvo JP: MRI protocol technique in the optimal therapeutic strategy of non-functioning pituitary adenomas. Eur J Endocrinol 2002;146:179–186.

17 Wolfsberger S, Ba-Ssalamah A, Pinker K, Mlynárik V, Czech T, Knosp E, trattnig S: Application of three-tesla magnetic resonance imaging for diagnosis and surgery of sellar lesions. J Neurosurg 2004;100:278–286.

18 Yoneoka Y, Watanabe N, Matsuzawa H, Tsumanuma I, Ueki S, Nakada T, Fujii Y: Preoperative depiction of cavernous sinus invasion by pituitary macroadenoma using three-dimensional anisotropy contrast periodically rotated overlapping parallel lines with reconstruction imaging on a 3-tesla system. J Neurosurg 2008;108:37–41.

19 Zimmerman RA: Imaging of intrasellar, suprasellar and parasellar tumors. Semin Roentgenol 1990;25:174–197.

20 Zirkzee EJM, Corssmit EPM, Biermasz NR, Brouwer PA, Wiggers-de-Bruine FT, Kroft LJM, van Buchem MA, Roelfsema F, Pereira AM, Smit JWA, Romijn JA: Pituitary magnetic resonance imaging is not required in the postoperative follow-up of acromegalic patients with long-term biochemical cure after transsphenoidal surgery. J Clin Endocrinol Metab 2004;89:4320–4324.

Prof. Dr. Michael Buchfelder
Department of Neurosurgery, University of Erlangen-Nürnberg
Schwabachanlage 6
D–91054 Erlangen (Germany)
Tel. +49 9131 85 34566, Fax +49 9131 85 34476, E-Mail nch-sekretariat@uk-erlangen.de

Pathogenesis of Familial Acromegaly

Mônica R. Gadelha[a,b] · Lawrence A. Frohman[c]

[a]Division of Endocrinology, Department of Internal Medicine, Hospital Universitário Clementino Fraga Filho, Universidade Federal do Rio de Janeiro, and [b]Instituto Estadual de Diabetes e Endocrinologia Luiz Capriglione, Rio de Janeiro, Brazil; [c]Section of Endocrinology, Department of Medicine, University of Illinois at Chicago, Chicago, Ill., USA

Abstract

Familial acromegaly may occur as a component of syndromes of multiple endocrine neoplasia or as isolated familial somatotropinoma (IFS), which is included in the spectrum of familial isolated pituitary adenoma (FIPA). We review the pathogenesis of IFS, from the detection of loss of heterozygosity at chromosome 11q13 and establishment of linkage to this chromosome region to the description of germline mutations in the aryl hydrocarbon receptor-interacting protein *(AIP)* gene. Approximately 40% of IFS families harbor an *AIP* mutation. In addition, we summarize the clinical features of IFS families with *AIP* mutations: The adenomas are diagnosed at a young age and are larger than in IFS patients without *AIP* mutations or in sporadic somatotropinomas, indicating more aggressive disease.

Copyright © 2010 S. Karger AG, Basel

Somatotropinomas occur with an annual incidence of three cases per million and a prevalence of 40–60 cases per million [1, 2]. Although the majority of GH-secreting adenomas are sporadic, a small number of somatotropinomas occur with a familial aggregation as a component of the classic syndromes of multiple endocrine neoplasia type 1 (MEN-1) and Carney complex (CNC) [3]. The MEN-1 syndrome is inherited in an autosomal dominant pattern with a high penetrance and is caused by a mutation in the tumor suppressor *MEN-1* gene that is located on chromosome region 11q13. GH-secreting adenomas occur in approximately 6% of patients with MEN-1. Most of these adenomas are associated with acromegaly, rather than gigantism, with the diagnosis being made between 30 and 50 years of age. The clinical characteristics of acromegaly in MEN-1 patients are similar to those described for sporadic acromegaly [3]. The CNC exhibits an autosomal dominant inheritance pattern and can be caused by a mutation in the tumor suppressor *PRKAR1A* (type 1 alpha regulatory subunit of protein kinase A) gene located on chromosome region 17q22–24 or by a mutation in a yet unidentified gene located at chromosome region 2p16 [4, 5]. Somatotropinomas are present in approximately 10% of patients with the CNC and can be associated with gigantism or acromegaly [3]. In 2006, patients with the MEN-1 phenotype but no *MEN-1* gene mutation have been described to harbor a germline mutation in the tumor suppressor $p27^{Kip1}$ gene, and this condition has been called multiple endocrine neoplasia type 4 (MEN-4) [6–8].

Familial acromegaly can also occur as isolated familial GH-secreting adenomas. This condition is designated isolated familial somatotropinoma (IFS) and is included in the syndrome of familial isolated pituitary adenoma (FIPA) [9, 10]. Isolated familial somatotropinoma is, therefore, defined as the occurrence of at least two cases of acromegaly or gigantism in a family that does not exhibit MEN-1, MEN-4 or CNC. The transmission pattern of IFS is autosomal-dominant with incomplete penetrance and, to date, approximately 90 IFS families have been reported in the literature [3, 11–13]. In around 70% of patients with IFS, the diagnosis occurs before the age of 30 years [12]. This young age is similar to that described for GH-secreting pituitary adenomas associated with CNC, but is in contrast to sporadic and MEN-1-associated somatotropinomas, which typically occurs between 30 and 50 years of age. The vast majority of IFS adenomas are macroadenomas that exhibit extra-sellar extension [3].

Isolated Familial Somatotropinoma: From Loss of Heterozygosity at Chromosome Region 11q13 to *AIP* Gene Mutation

In 1999, we described an IFS family from Brazil that consisted of 13 siblings, 6 of whom were affected [9]. In addition, the father's brother died at the age of 18 years and was suspected of having the disease. All affected members were diagnosed at a young age, between 13 and 24 years, and 5 of the 6 had aggressive octreotide-resistant macroadenomas. Molecular analysis revealed loss of heterozygosity on chromosome region 11q13 (*MEN-1* locus) [9], which was previously reported by Yamada et al. [14] in a Japanese IFS family. However, neither a *MEN-1* mutation nor decreased gene expression were detected [9]. Subsequently, tumor deletion mapping and haplotype analyses in the Brazilian family established linkage to chromosome region 11q13 and identified the *IFS* tumor suppressor gene candidate interval (fig. 1). In addition, a potential second locus at chromosome region 2p16–12 was identified [15]. These data suggested that a tumor suppressor gene located at chromosome region 11q13 but distinct from the *MEN-1* gene was involved in the pathogenesis of IFS.

Recently, Vierimaa et al. [16] studied two Finnish families with FIPA using chip-based technologies and genealogic data. One family consisted of patients with somatotropinomas and prolactinomas, but only the individuals with acromegaly or gigantism were considered affected for the linkage analysis. The other family exhibited IFS. The authors found germline mutations in a gene located at chromosome region 11q13 in the *IFS* candidate interval, the aryl hydrocarbon receptor (AhR)-interacting protein *(AIP)* gene (fig. 1). Mutation analysis of the Brazilian family revealed a germline *AIP* stop codon mutation (E24X) within exon one (fig. 2) [12]. Because this mutation

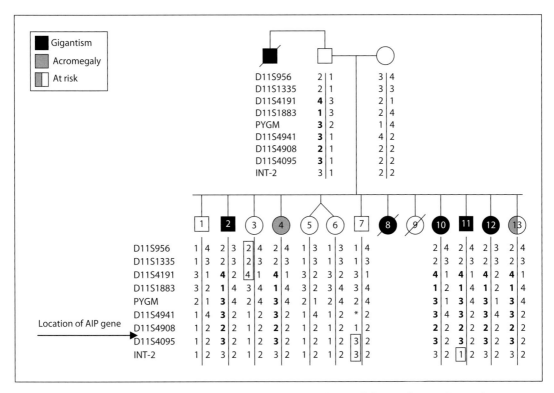

Fig. 1. Haplotype analysis of the Brazilian family. Numbers indicate allele size for 9 polymorphic microsatellite markers located at chromosome 11q12.3–13.3. Open rectangles represent meiotic recombination events. The markers that are included in the candidate interval are shown in bold (the centromeric limit of the interval was established with allelotype analysis and the telomeric limit by the meiotic recombination event present in the haplotype of individual 11 [15]). The location of the *AIP* gene is shown. The asterisk denotes an uncommon allele size.

occurs in the 5′ region of the gene, the AIP protein is predicted not to be generated. Approximately 40% of IFS families studied were shown to have an *AIP* mutation [12]. Therefore, another gene is involved in the pathogenesis of the *AIP* negative IFS families or these families harbor large AIP genomic rearrangements, intronic mutations leading to abnormal splicing of the exons or promoter mutations. In addition, no somatic AIP mutation has been found in sporadic somatotropinomas [12], supporting the involvement of other gene(s) in the pathogenesis of GH-secreting adenomas.

The Aryl Hydrocarbon Receptor Interacting Protein *(AIP)* Gene

The *AIP* gene is located in chromosome region 11q13, contains six exons, and encodes a 330 amino acid chaperone protein. Leontiou et al. [12] demonstrated that the *AIP* gene has properties consistent with a tumor suppressor gene since overexpression of the

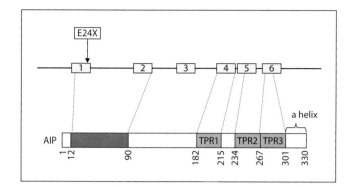

Fig. 2. The *AIP* gene and the mutation found in the Brazilian family. The structure of the AIP protein is shown at the bottom.

wild-type AIP decreased cell proliferation in various cell lines. In addition, the authors showed that the protein derived from the mutant *AIP* genes completely or partially loses this function. Among the several possible interacting partners of AIP are AhR and phosphodiesterase (PDE) isoforms [17, 18]. However, to date, the exact mechanism by which this chaperone protein causes tumor suppression remains unknown. AIP has both stimulating and inhibiting effects on the *AhR* gene, but the importance of these actions in the process of tumor formation requires further investigation. Alterations in the cAMP pathway seem to be involved in somatotropinoma pathogenesis: Carney complex patients have an inactivating mutation in the *PRKAR1A* gene and up to 40% of sporadic GH-secreting adenomas have a *gsp*-activating mutation [4, 19, 20]. In addition, Leontiou et al. [12] reported that all *AIP* mutations that they studied led to a disruption of the interaction between AIP and PDE4A5. Therefore, the AIP interaction with PDE may be involved in the control of cell proliferation exerted by AIP.

In the normal pituitary, AIP is present only in GH and prolactin cells in which it is associated with secretory granules. In contrast, AIP is expressed in all types of sporadic pituitary adenomas. However, AIP is normally colocalized only in sporadic somatotropinomas and not in prolactin-, ACTH- and FSH-secreting tumors [12]. Based on these findings, we can speculate that the AIP tumor suppressor function is primarily of importance for GH-secreting cells. In prolactinomas, Cushing's disease and nonfunctioning adenomas, the AIP expression may represent a secondary phenomenon, a mechanism turned on to control tumor growth. In support of this hypothesis is the observation that germline *AIP* mutations occur mainly in typical IFS families or in FIPA families in which at least one member has a somatotroph adenoma.

Clinical Features of Families with *AIP* Mutations

Adenomas from IFS families containing an AIP mutation are diagnosed at a young age; approximately 70% being diagnosed by the age of 25 years. The tumors are larger than in IFS patients without AIP mutations or in sporadic somatotropinomas, indicating more aggressive disease [12]. The Brazilian IFS family, in which we established linkage to the *AIP* gene region, may be considered a paradigm of a family containing an *AIP* mutation [15]: The index case was diagnosed at 24 years, had a huge macroadenoma that was treated by surgery, radiotherapy and somatostatin analogs. The other affected members were diagnosed at ages between 13–18 years and, except for one patient who was diagnosed during genetic screening, all had aggressive disease. These patients were resistant to treatment with octreotide; although whether this is a characteristic of *AIP* positive patients as a group requires further investigation.

Conclusion

The spectrum of familial acromegaly has increased over the past decade to include syndromes associated with multiple endocrine tumors (MEN-1, MEN-4, and CNC) and FIPA of which IFS constitute the largest subgroup. Although mutations in the *AIP* gene have been found in 40% of families with IFS, the genetic abnormalities in the remaining families are yet to be identified as are their relevance to sporadic somatotropinomas.

References

1. Alexander L, Appleton D, Hall R, Ross WM, Wilkinson R: Epidemiology of acromegaly in Newcastle region. Clin Endocrinol (Oxf) 1980;12: 71–79.
2. Etxabe J, Gaztambide P, Latorre P, Vasquez JA: Acromegaly: an epidemiological study. J Endocrinol Invest 1993;16:181–187.
3. Gadelha MR, Kineman RD, Frohman LA: Familial somatotropinomas: clinical and genetic aspects. Endocrinologist 1999;9:277–285.
4. Kirschner LS, Carney JA, Pack SD, Taymans SE, Giatzakis C, Cho YS, Cho-Chung YS, Stratakis CA: Mutations of the gene encoding the protein kinase A type I-alpha regulatory subunit in patients with the Carney complex. Nat Genet 2000;26: 89–92.
5. Stratakis CA, Carney JA, Lin JP, Papanicolaou DA, Karl M, Kastner DL, Pras E, Chrousos GP: Carney complex, a familial multiple neoplasia and lentiginosis syndrome: analysis of 11 kindreds and linkage to the short arm of chromosome 2. J Clin Invest 2006;97:699–705.
6. Pellegata NS, Quintanilla-Martinez L, Siggelkow H, Samson E, Bink K, Hofler H, Fend F, Graw J, Atkinson MJ: Germ-line mutations in p27Kip1 cause a multiple endocrine neoplasia syndrome in rats and humans. Proc Natl Acad Sci USA 2006; 103:15558–15563.
7. Owens M, Stals K, Ellard S, Vaidya B: Germline mutations in the CDKN1B gene encoding p27Kip1 are a rare cause of multiple endocrine neoplasia type 1. Clin Endocrinol (Oxf) 2009;70:499–500.
8. Igreja S, Chahal HS, Akker SA, Gueorguiev M, Popovic V, Damjanovic S, Wass JA, Quinton R, Grossman AB, Korbonits M: Assessment of p27 (cyclin-dependent kinase inhibitor 1B) and AIP (aryl hydrocarbon receptor-interacting protein) genes in MEN1 syndrome patients without any detectable MEN1 gene mutations. Clin Endocrinol (Oxf) 2009;70:259–264.
9. Gadelha MR, Prezant TR, Une KN, Glick RP, Moskal SF, Vaisman M, Melmed S, Kineman RD, Frohman LA: Loss of heterozygosity on chromosome 11q13 in two families with acromegaly/gigantism is independent of mutations of the multiple endocrine neoplasia type I gene. J Clin Endocrinol Metab 1999;84:249–256.

10 Daly AF, Jaffrain-Rea ML, Ciccarelli A, et al: Clinical characterization of familial isolated pituitary adenomas. J Clin Endocrinol Metab 2006;91:3316–3323.

11 Daly AF, Vanbellinghen JF, Khoo SK, et al: Aryl hydrocarbon receptor-interacting protein gene mutations in familial isolated pituitary adenomas: analysis in 73 families. J Clin Endocrinol Metab 2007;92:1891–1896.

12 Leontiou CA, Gueorguiev M, van der SJ, et al: The role of the AIP gene in familial and sporadic pituitary adenomas. J Clin Endocrinol Metab 2008;93: 2390–2401.

13 Georgitsi M, Heliovaara E, Paschke R, et al: Large Genomic Deletions in AIP in Pituitary Adenoma Predisposition. J Clin Endocrinol Metab 2008;93: 4146–4151.

14 Yamada S, Yoshimoto K, Sano T, Takada K, Itakura M, Usui M, Teramoto A: Inactivation of the tumor suppressor gene on 11q13 in brothers with familial acrogigantism without multiple endocrine neoplasia type 1. J Clin Endocrinol Metab 1997;82:239–242.

15 Gadelha MR, Une KN, Rohde K, Vaisman M, Kineman RD, Frohman LA: Isolated familial somatotropinomas: establishment of linkage to chromosome 11q13.1–11q13.3 and evidence for a potential second locus at chromosome 2p16-12. J Clin Endocrinol Metab 2000;85:707–714.

16 Vierimaa O, Georgitsi M, Lehtonen R, Vahteristo P, Kokko A, Raitila A, Tuppurainen K, Ebeling TM, Salmela PI, Paschke R, Gundogdu S, De ME, Makinen MJ, Launonen V, Karhu A, Aaltonen LA: Pituitary adenoma predisposition caused by germline mutations in the AIP gene. Science 2006;312: 1228–1230.

17 Carver LA, Bradfield CA: Ligand-dependent interaction of the aryl hydrocarbon receptor with a novel immunophilin homolog in vivo. J Biol Chem 1997; 272:11452–11456.

18 Bolger GB, Peden AH, Steele MR, MacKenzie C, McEwan DG, Wallace DA, Huston E, Baillie GS, Houslay MD: Attenuation of the activity of the cAMP-specific phosphodiesterase PDE4A5 by interaction with the immunophilin XAP2. J Biol Chem 2003;278:33351–33363.

19 Donangelo I, Gadelha M: Bases moleculares dos adenomas hipofisários com ênfase nos somatotropinomas. Arq Bras Endocrinol Metab 2004;48:464–479.

20 Taboada GF, Tabet ALO, Naves LA, Carvalho DP, Gadelha MR: Prevalence of gsp oncogene in somatotropinomas and clinically non-functioning pituitary adenomas: our experience. Pituitary 2009;12: 165–169.

Mônica R. Gadelha
Hospital Universitário Clementino Fraga Filho
Rua Professor Rodolpho Paulo Rocco, 255, 9E23 – Cidade Universitária – Ilha do Fundão –
Rio de Janeiro CEP 21941–913 (Brasil)
Tel. +55 21 2562 2323, Fax +55 21 2562 2111, E-Mail mgadelha@hucff.ufrj.br

Functional Role of the RET Dependence Receptor, GFRa Co-Receptors and Ligands in the Pituitary

Montserrat Garcia-Lavandeira · Esther Diaz-Rodriguez · Maria E.R. Garcia-Rendueles · Joana S. Rodrigues · Sihara Perez-Romero · Susana B. Bravo · Clara V. Alvarez

Department of Physiology, School of Medicine, IDIS University of Santiago de Compostela (USC), Santiago de Compostela, Spain

Abstract

The RET receptor is a tyrosine kinase receptor implicated in kidney and neural development. In the adenopituitary RET and the co-receptor GFRa1 are expressed exclusively in the somatotrophs secreting GH. RET is implicated in a clever pathway to maintain at physiological levels the number of somatotrophs and the GH production. Thus, in absence of its ligand GDNF, RET induces apoptosis through massive expression of Pit-1 leading to p53 accumulation. In the presence of the ligand GDNF, RET activates its tyrosine kinase and promotes survival at the expense of reducing Pit-1 expression and downregulating GH. Recent data suggest that RET can also have a second role in pituitary plasticity through a second co-receptor GFRa2.

Copyright © 2010 S. Karger AG, Basel

The pituitary is an endocrine gland in which specialized cells secrete hormones. In everyday life and also to adapt to puberty, pregnancy or hypothyroidism, the pituitary must regulate specifically and independently each secretory cell-type, not only through the release of hormones but also by regulating cell numbers [1]. Recent data implicate the RET receptor as one of the players in the control of such plasticity in the pituitary.

RET Receptor, Its Co-Receptors and Its Ligands

The RET receptor is a membrane receptor having an extracellular, a transmembrane and a cytoplasmic domain. Although originally discovered as an oncogene in a transfected cell-line [2], RET was soon detected as an endogenous gene, strongly expressed in the central nervous system (CNS), lymphatic organs and testis [3, 4]. RET is a long protein, expressed as two isoforms due to alternative splicing [5]. From the common glycine 1063 in the C-terminal cytoplasmic domain, the RET long isoform (RET-L) contains 51 amino acids while the RET short isoform (RET-S) has only 9 amino acids. Since its discovery, RET has been considered a classical tyrosine-kinase receptor that upon binding to its ligand, GDNF, cross-dimerizes and autophosphorylates some tyrosines (Y) in the C-terminal tail [6–8], thus attracting and activating other target proteins. In order to bind GDNF, RET needs a co-receptor known as GFRa1. This co-receptor is a protein bound to the extracellular side of the plasma membrane through a glycosyl-phosphatidyl-inositol (GPI) anchor that helps RET to bind the ligand. A surprising further discovery was that of three other ligands and co-receptors in addition to GDNF/GFRa1 that are able to activate RET: neurturin (NTN) and GFRa2, artemin (ART) and GFRa3 and persephin (PSP) and GFRa4 (reviewed in [9]) (fig. 1). As a consequence, a complex puzzle has begun to be revealed in which RET activates different transduction pathways, depending partly on the ligand/GFRa that it is bound to but mainly on the cell-type involved. Although much work remains in order to fully characterize the RET system, one recurrent pathway activated by GFRa/RET is the PI3K/AKT complex [10].

The function of the normal RET receptor has been intensely studied in neurons and in kidney development. In relation to human pathology, sporadic RET translocations have been implicated in the etiology of papillary thyroid carcinoma (RET/PTC). At least ten RET/PTC variants have been described in which the RET tyrosine-kinase domain is fused to other genes, although only RET/PTC1 (fused to the 5′ end of the CCDC6 gene) and RET/PTC3 (fused with the NCOA4 gene) are common [11, 12].

More than 110 congenital RET mutations are also well known for causing two diseases in humans: Hirschprung's disease and MEN2. In Hirschprung's disease, or colonic aganglionosis, mutations are considered as inactivating RET function while in the MEN2 variants MEN2A, MEN2B and FMTC mutations are considered as activating RET function. However, it is not well explained how certain mutations can cause both Hirschprung's disease and MEN2 [13, 14].

GDNF/GFRa1/RET System in the Pituitary

Some years ago mutations in RET or GDNF were excluded as a cause of pituitary adenomas [15]. At that time, however, there had not been any studies of the expression of these genes in the normal pituitary. Our group described for the first time the expression patterns of RET, GFRa1 and GDNF in the rat pituitary, at both mRNA and protein levels [16]. Using double-immunofluorescence, an interesting observation was made: the only cells in the adenopituitary expressing the receptors GFRa1/RET were the somatotrophs, all other secretory cell types being negative [16]. Similar

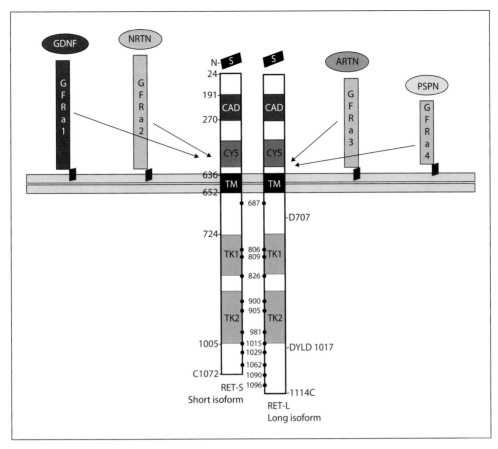

Fig. 1. The RET/GFRa system. The tyrosine-kinase receptor RET has an extracellular region with consensus cadherin-type domain, and a cysteine-rich domain. After the transmembrane domain (TM), the cytoplasmic tail contains the tyrosine-kinase domain. In the presence of any of its four GFRa extracellular co-receptors (GFRa 1–4) and ligands (GDNF, NTN, ART and PSP), RET crossdimerize activating its tyrosine-kinase activity. The main tyrosines phosphorylated upon activation are shown. Also shown are the key aspartic (D) amino acids in the two putative consensus caspase-sensitive cleavage domains found within the cytoplasmic tail.

studies in human pituitary have shown the same result. The GDNF/GFRa1/RET system has also been shown to be expressed in the adenopituitary and again specifically in somatotrophs [17] (fig. 2).

Two questions arose directly from those expression studies: what is the role of RET in the somatotrophs, and what conserved mechanism(s) underly the distinctive expression of RET in mammalian pituitary somatotrophs? To study the function of RET in somatotrophs, we began by using a more simple system than that of the whole organ. Some rat pituitary cell-lines maintain the ability to synthesize and secrete hormones, even though they have lost some important receptors for hypothalamic factors [18]. We used the GH4C1 cell line which maintains the ability to secrete

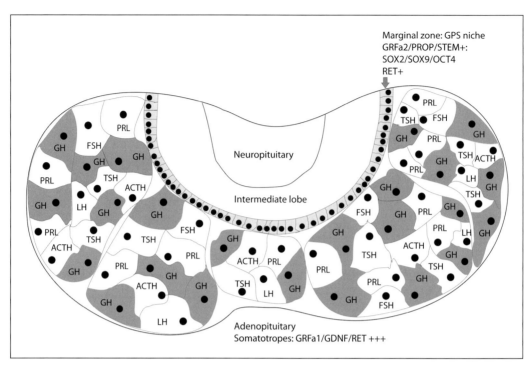

Fig. 2. Expression of RET in the pituitary. Somatotrophs, but not other secretory cells, express RET and GFRa1 (dark grey). In the marginal zone between Adenopituitary and Intermediate lobe the GPS niche of stem cells express RET although less intense (light grey). These cells do not express GFRa1 but instead they are intensively expressing the GFRa2 co-receptor in addition to stem markers such as Prop1, Oct4, Sox2 and Sox9.

GH and express GHSR receptors although it does not express GHRH receptors, and furthermore, while GH4C1 cells express GDNF and GFRa1, they do not express the RET receptor [19]. Although in transient transfected GH4C1 cells we could detect RET expression, we were unable to obtain a GH4C1 population that stably expressed RET. Through detailed studies we were able to demonstrate a strong apoptotic action of RET in GH4C1 somatotrophs that was blocked by the ligand GDNF, inducing survival. Apoptosis was induced by both isoforms RET-L and RET-S, while GDNF inhibited cell death in the presence of either receptor [19].

RET as a Member of the Family of Dependence Receptors

Some years ago a new type of receptor was described. These receptors are processed in the absence of their ligand and induce cell-death through apoptotic pathways. In the presence of the ligand the receptor remains whole and becomes activated in its classical transduction pathways, phosphorylating or activating other intracellular proteins.

Such receptors were called 'dependence receptors' since the cells in which they were expressed were absolutely dependent on the presence of their ligand for their survival (reviewed in [20–22]). Members of this family are the DCC and UNC5H2 receptors for netrin I, some integrins, the PTC receptor for SHH and the androgen receptor. During the last years, there was a suggestion that RET could be acting as a dependence receptor in some cell-lines [23].

To Die of Success: RET Kills through PIT Overexpression

Our data in GH4C1 somatotrophs, where RET induced apoptosis while GDNF induced survival, placed RET as a dependence receptor in somatotrophs but did not explain the mechanism. In parallel, we studied the expression of genes phenotypically characteristic of somatotrophs such as PIT (POU1F1, PIT-1). RET transfection induced a marked and sustained increase in PIT both at the mRNA and protein levels, and GDNF also blocked this action [19]. The tyrosine-kinase activity of RET was not required for these actions. When the same cells were transfected with a kinase-dead mutant receptor, apoptosis or PIT induction were still induced. However, GDNF failed to induce survival in the presence of this mutant, demonstrating the requirement of the tyrosine-kinase pathway to survival.

One property common to all dependence receptors is their processing by intracellular caspases [20]. In GH4C1, RET presence was positively correlated with caspase-3 activation, while a RET point mutant of one of the two caspase-3 putative consensus sites could not induce apoptosis. Surprisingly, however, PIT induction was also blocked. Indeed, PIT induction by normal RET was blocked by caspase-3 inhibitors. Through a series of experiments, the data pointed to a relationship between strong PIT induction and apoptosis. Finally, we compared transfection of PIT with RET and with both PIT and RET together. All three experiments gave comparable levels of apoptosis without any additive effects. In fact, RET-induced PIT overexpression led to p53 accumulation and cell death [19]. Opposite PIT siRNA blocked either PIT or RET-induced p53 expression and apoptosis, demonstrating a direct relationship between both proteins.

We then tried to elucidate the cytoplasmic pathway from RET to PIT. Through a series of complex co-immunoprecipitations, we demonstrated that RET forms an intracellular complex with caspase-3 and the protein kinase PKC-delta (PKCd) on the cytoplasmic surface of the plasma membrane. Either caspase-3 or PKCd need to be cleaved and phosphorylated to become activated [24–26]. In one previous publication, caspase-3 and PKCd had been found to form a complex [27]. In somatotrophs, the presence of the intracellular tail of undimerized RET initiates both the formation of the complex itself and the proteolytic processing of the three proteins that generates an 18-kDa caspase-3 (full length 30 kDa), a 40-kDa PKC-d (full length 80 kDa) and a 55-kDa intracellular RET fragment (IC-RET; RET full length is 150–170 kDa).

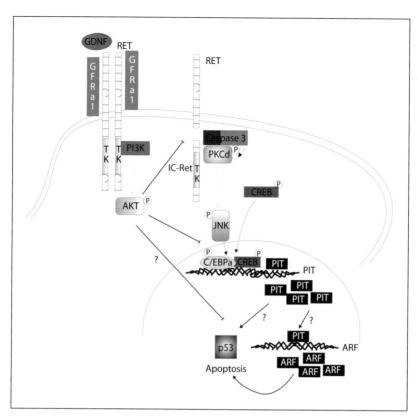

Fig. 3. The RET/PIT/p53 pathway opposite the RET/GFRa1/GDNF pathway in the somatotroph cell. Caspase-3 processed IC-Ret regulates somatotroph differentiation by potently inducing Pit-1 expression. Ret induced Pit-1 overexpression, leads to increased p53 expression and apoptosis. The Pit-1 overexpression is mediated by sustained activation of PKCδ, JNK, C/EBPa and CREB. In the presence of GDNF, however, Akt is activated and the RET-processing, the Pit-1 over-expression, and consequent apoptosis, is blocked.

Immediately afterwards, two other proteins become phosphorylated and activated, the kinase JNK and the transcription factor CREB. Both phosphorylations depend on PKCd phosphorylation to become activated since rottlerin, a specific PKCd inhibitor, prevents JNK or CREB phosphorylation, PIT induction and apoptosis [19]. On the other hand, addition of GDNF causes dimerization and activation of RET tyrosine-kinase, preventing intracellular processing of the receptor – or indeed of other proteins – and leading to AKT phosphorylation, thus blocking apoptosis and leading to cell survival. Inhibitors of p-AKT, even in the presence of GDNF, again lead to apoptosis (fig. 3).

Previously, our group had demonstrated an induction of the PIT expression at the promoter level by GHRH [28] or by ghrelin [29]. Both effects were transient and rapid, lasting not more than two hours, and were exerted at the two CRE-response elements (around -200 bp) on the PIT promoter. These results raised many questions since the tandem CRE-sites exist only in the murine PIT promoter and are not conserved in the human PIT promoter. We now understand the importance of GHRH or ghrelin activation of the PIT promoter for a rapid response, since a more chronic PIT induction leads to apoptosis. In fact, both GHRH and ghrelin were able to activate p-AKT at the same time as activating the cAMP/CREB pathway. We now think that this ability is the key factor in the swiftness of GHRH/ghrelin action at the PIT promoter, since GDNF, the ligand inducing RET dimerization and tyrosine-kinase activity, inhibits PIT transcription through p-AKT.

When we started to study PIT-promoter activation via RET we thought that the tandem CRE region would also be implicated, since RET induced CREB phosphorylation. However, the promoter region implicated in the PIT induction was an element at −300 bp. This new region was a consensus site for c/EBP alpha (c/EBPa), a leucine zipper transcription factor of fundamental importance during embryonic development [30–33]. In vitro studies related c/EBPa in the pituitary with transcriptional activation of PRL and GH genes [34–37]. Importantly, the c/EBPa site is conserved in both murine and human PIT promoters. Using chromatin immunoprecipitation studies, we demonstrated the specific binding of c/EBPa and CREB to this region of the PIT promoter after RET transfection, which was blocked by GDNF, caspase-3 inhibitors, rottlerin or the dominant-negative mutant killer-CREB [19] (fig. 3).

Importance of the RET Pathway in vivo: RET KO Pituitary and Prevention of Tumor Growth

Our results in the GH4C1 cell line were easily reproduced in primary pituitary cultures with the advantage that the endogenous receptor was present. In these cultures more than 60% of the cells are somatotrophs. Immediately after replacing the culture medium with a low serum medium, an immediate processing of the endogenous RET receptor was observed, with caspase-3/PKCd activation and strong PIT induction leading to massive apoptosis [19]. All these events were prevented by the addition of GDNF which induced RET tyrosine-kinase activity and p-AKT activation.

The RET KO mouse has a strong phenotype with profound alterations in the kidneys and reproductive system. Moreover, animals die within a few hours of birth due to failure in peristalsis and digestion due to the absence of parasympathetic innervation in the intestines [38]. Depending on how important the RET pathway is in pituitary embryonic development relative to its importance in pituitary maintenance after birth, we could predict a RET KO phenotype. Therefore, we characterized the pituitaries of wild-type and RET KO newborn mice. The pituitaries of the RET KO were bigger both in total volume and surface area per section. But this enlargement was entirely within the adenopituitary (AP), with the intermediate lobe and neuropituitary showing no difference in size compared with the wild-type pituitary. Proliferative activity was also no different in the two mice genotypes. When all pituitary secretory-cell types were analyzed, PIT-expressing cells were found in significantly increased numbers in the RET KO pituitaries, while SF-1 (gonadotroph) cells were identical in number in the KO as in the wild type. Finally, somatotrophs were the only Pit1+ secretory cell-type that increased in number, with no differences being found for either lactotrophs or thyrotrophs [19]. Interestingly, c/EBPa KO mice, like RET KO mice, also die after birth of hypoglycemia and lung deficiencies [32, 33]. It will be important to investigate in the future whether or not this KO also has a pituitary phenotype.

Although somatotrophs are the most abundant cells in the pituitary, somatotrophic tumors are the least frequent. Furthermore, in aggressive adenomas and pituitary carcinomas, somatotroph adenomas are an uncommon source. We wanted to investigate whether the RET/PIT pathway was protecting somatotrophs against tumor growth, and if so, whether we could utilise this system to prevent growth in other pituitary tumors. Since there is not a good inducible model for somatotrophic tumors, we chose to induce a lactotroph adenoma through repeated estrogen injections. Lactotrophs do not express RET but they are PIT-expressing. We prepared retroviral particles bearing RET and control particles with an empty virus. We chose retroviruses as carriers of RET because they primarily infect proliferating cells and not resting differentiated cells, and would therefore specifically infect tumor cells. Using stereotactic techniques we injected the retrovirus precisely into the adenopituitary and waited a week. As expected, the pituitaries treated with estrogens and injected with control retrovirus weighed twice the amount of normal untreated pituitaries, with a high level of statistical significance. However, pituitaries treated with estrogens but injected with RET-expressing retrovirus showed no difference in weight compared with normal non-injected untreated pituitaries. Histological analysis showed lactotrophs expressing the RET receptor to be present in pituitaries treated with estrogens and infected with RET-expressing retroviruses, but not in normal pituitaries nor in pituitaries treated with estrogen but infected with empty viruses. Moreover, in extracts of pituitaries infected with the RET-expressing virus, but not in extracts of the empty-virus infected pituitaries, the presence of fragmented IC-RET, together with high levels of p-CREB and p-JNK, PIT and p53 were detected. RET infection was also accompanied by a strong tendency to apoptosis as detected through PARP cleavage [19]. Taken together, these data from a model of pituitary tumorogenesis strongly suggest the potential usefulness of the RET/PIT pathway in designing new drugs for pituitary tumor treatment (fig. 4).

More Questions for Future Answers

The RET/PIT/p53 pathway is a relatively well understood mechanism for the precise control of the number and function of somatotrophs. There are, however, still some particular questions that need to be answered. The first is the exact nature of the IC-RET/PKCd/caspase-3 complex. Is caspase-3 only acting as a protease, processing/activating both RET and PKCd to phosphorylate JNK/CREB? Or may the opposite

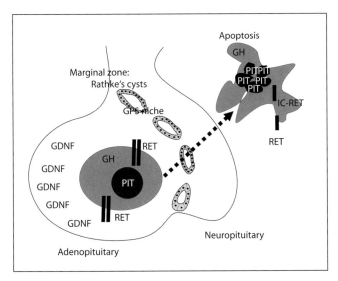

Fig. 4. Importance of the pathway to prevent somatotroph tumors in humans. Somatotrophs are under strong pressure to stay in the pituitary. If a precursor cell is to become a somatrotroph, a GH-secreting cell, it needs Pit-1, and to maintain sufficient Pit-1 expression after birth, Ret expression is required. But if the somatotroph migrates outside the pituitary, to tissues in which GDNF is present at much lower concentrations, Ret-induced Pit-1 overexpression will kill it. This would further explain why all GH-secreting adenomas maintain normal Ret/GFRa1 and GDNF expression, and why GH-secreting adenomas are incapable of metastasis. The human pituitary has also a GPS niche in the marginal zone, where progenitors expressed similar stem-cell markers and RET as the murine GPS niche.

be true: is IC-RET activating caspase-3, such that, instead of processing RET/PKCd, it would be able to activate other apoptotic pathways? Our results strongly suggest that the first is true, but more detailed biochemical studies are needed to precisely elucidate the nature of the ternary complex. A second important point would be to determine at the DNA level the exact contribution of the two transcription factors implicated in the PIT activation, c/EBPa and CREB: whether they are pre-assembled, bind upon activation or contribute independently to PIT activation. Another area for future study is to investigate how PIT can lead to stabilization of p53, and especially whether this is a direct effect or mediated by other known or unknown genes.

It is also important that the RET/PIT/p53 pathway has been conserved through evolution. Our data appear to be common in the rodent and human pituitary. From studies in zebrafish, birds, rodents, cattle, monkeys and humans, it has also been demonstrated that the RET tyrosine-kinase receptor is one of the most evolutionarily conserved receptors, having an average of only 1.3 substitutions per position per 100 million years [39]. The intracellular region IC-RET is, furthermore, the most conserved part of the receptor as a whole. More studies of the evolutionary aspects of the pituitary are needed, and it is encouraging for future work that RET is expressed in the pituitary of zebrafish [40].

It will also be interesting to elucidate pathways used by other secretory cell-types in the pituitary. The fact that RET is not expressed in non-somatotroph cell types precludes its involvement, but other members of the family of dependence receptors could still be regulating survival and function of lactotrophs, thyrotrophs, corticotrophs or gonadotrophs. Alternatively, perhaps only the somatotrophs need such specific regulation due to the nature of the hormone they secrete acting in the body as a whole rather than on a specific target organ. Thus the RET/GFRa1/PIT system would help to prevent somatotroph tumors. The somatotroph needs RET to maintain the PIT and GH expression over a longer period. More GDNF within the pituitary would result in more somatotrophs surviving but less PIT and less GH being expressed per cell. Any somatotroph leaving pituitary areas rich in GDNF would in effect be committing suicide. The data obtained in human pituitary adenomas fit with this hypothesis. In our experimental series, all somatotroph adenomas tested maintained their expression of RET/GFRa1 and GDNF [17]. Furthermore, it is known that somatotroph adenomas are less frequent than other types of adenomas, in spite of somatotrophs making up more than half the total number of secretory cells in the normal pituitary. Coincidentally, in a recent review on human aggressive pituitary tumors and carcinomas [Colao et al., this vol., pp. 94–108], the tumors least likely to have an aggressive phenotype were somatotroph tumors. This fact also highlights the likely importance of the RET/GFRa1/PIT pathway to pituitary tumor treatment or diagnosis in the future.

Finally, it should be remembered that RET has three other ligands and GFRa co-receptors. Some of these are expressed on the adenopituitary [unpubl. data from our group]. Additionally, our group has recently demonstrated the existence of a niche of stem cells, conserved in the murine and human pituitary. We have called this the GPS niche since the small cells within it express the RET co-receptor GFRa2, the pituitary transcription factor Prop1 and stem cell markers such as Sox2, Sox9 and Oct4 [41]. These cells are also weakly positive for RET. GPS cells have stem cell characteristics both in vivo and in vitro, and some animal models of hypopituitarism due to alteration of cell cycle proteins [42] present a phenotype in the GPS niche [41]. The relationship between cell-cycle proteins and stem cells of the pituitary has been recently started to be unveiled [43]. Therefore, the study of RET/GFRa2 in the GPS niche and its importance to the pituitary physiology will make for an interesting future in this field.

Acknowledgements

We thank Carlos Dieguez for helpful scientific discussions and suggestions. We thank Pamela Lear for its expert reading and corrections of scientific English.

This work was funded by grants from the Xunta de Galicia PGIDIT 05BTF20803PR, 06PPXIB208107PR, 2006/PX259 and 2009/PX23; from the Ministry of Education and Science (MICINN) SAF2004–03131 and BFU2007–60571.

References

1 Melmed S: Mechanisms for pituitary tumorigenesis: the plastic pituitary. J Clin Invest 2003;112:1603–1618.
2 Takahashi M, Ritz J, Cooper GM: Activation of a novel human transforming gene, ret, by DNA rearrangement. Cell 1985;42:581–588.
3 Tahira T, Ishizaka Y, Sugimura T, Nagao M: Expression of proto-ret mRNA in embryonic and adult rat tissues. Biochem Biophys Res Commun 1988;153:1290–1295.
4 Takahashi M, Buma Y, Iwamoto T, Inaguma Y, Ikeda H, Hiai H: Cloning and expression of the ret proto-oncogene encoding a tyrosine kinase with two potential transmembrane domains. Oncogene 1988; 3:571–578.
5 Tahira T, Ishizaka Y, Itoh F, Sugimura T, Nagao M: Characterization of ret proto-oncogene mRNAs encoding two isoforms of the protein product in a human neuroblastoma cell line. Oncogene 1990;5: 97–102.
6 Takahashi M, Cooper GM: Ret Transforming Gene Encodes a Fusion Protein Homologous to Tyrosine Kinases. Mol Cell Biol 1987;7:1378–1385.
7 Jing S, Wen D, Yu Y, et al: GDNF-induced activation of the ret protein tyrosine kinase is mediated by GDNFR-alpha, a novel receptor for GDNF. Cell 1996;85:1113–1124.
8 Treanor JJ, Goodman L, de Sauvage F, et al: Characterization of a multicomponent receptor for GDNF. Nature 1996;382:80–83.
9 Airaksinen MS, Saarma M: The GDNF family: signalling, biological functions and therapeutic value. Nat Rev Neurosci 2002;3:383–394.
10 Santoro M, Carlomagno F, Melillo RM, Fusco A: Dysfunction of the RET receptor in human cancer. Cell Mol Life Sci 2004;61:2954–2964.
11 Pierotti MA, Santoro M, Jenkins RB, et al: Characterization of an inversion on the long arm of chromosome 10 juxtaposing D10S170 and RET and creating the oncogenic sequence RET/PTC. Proc Natl Acad Sci USA 1992;89:1616–1620.
12 Bongarzone I, Butti MG, Coronelli S, et al: Frequent activation of ret proto-oncogene by fusion with a new activating gene in papillary thyroid carcinomas. Cancer Res 1994;54:2979–2985.
13 Arighi E, Popsueva A, Degl'Innocenti D, et al: Biological effects of the dual phenotypic Janus mutation of ret cosegregating with both multiple endocrine neoplasia type 2 and Hirschsprung's disease. Mol Endocrinol 2004;18:1004–1017.
14 Plaza-Menacho I, Burzynski GM, de Groot JW, Eggen BJ, Hofstra RM: Current concepts in RET-related genetics, signaling and therapeutics. Trends Genet 2006;22:627–636.
15 Yoshimoto K, Tanaka C, Moritani M, et al: Infrequent detectable somatic mutations of the RET and glial cell line-derived neurotrophic factor (GDNF) genes in human pituitary adenomas. Endocr J 1999;46:199–207.
16 Urbano AG, Suarez-Penaranda JM, Dieguez C, Alvarez CV: GDNF and RET-gene expression in anterior pituitary-cell types. Endocrinology 2000; 141:1893–1896.
17 Japon MA, Urbano AG, Saez C, et al: Glial-derived neurotropic factor and RET gene expression in normal human anterior pituitary cell types and in pituitary tumors. J Clin Endocrinol Metab 2002;87: 1879–1884.
18 Coya R, Alvarez CV, Perez F, Gianzo C, Dieguez C: Effects of TGF-beta1 on prolactin synthesis and secretion: an in-vitro study. J Neuroendocrinol 1999;11:351–360.
19 Canibano C, Rodriguez NL, Saez C, et al: The dependence receptor Ret induces apoptosis in somatotrophs through a Pit-1/p53 pathway, preventing tumor growth. EMBO J 2007;26:2015–2028.
20 Mehlen P, Thibert C: Dependence receptors: between life and death. Cell Mol Life Sci 2004;61: 1854–1866.
21 Allouche M: ALK is a novel dependence receptor: potential implications in development and cancer. Cell Cycle 2007;6:1533–1538.
22 Bernet A, Fitamant J: Netrin-1 and its receptors in tumour growth promotion. Expert Opin Ther Targets 2008;12:995–1007.
23 Bordeaux MC, Forcet C, Granger L, et al: The RET proto-oncogene induces apoptosis: a novel mechanism for Hirschsprung disease. EMBO J 2000;19: 4056–4063.
24 Majumder PK, Pandey P, Sun X, et al: Mitochondrial translocation of protein kinase C delta in phorbol ester-induced cytochrome c release and apoptosis. J Biol Chem 2000;275:21793–21796.
25 Sun X, Wu F, Datta R, Kharbanda S, Kufe D: Interaction between protein kinase C delta and the c-Abl tyrosine kinase in the cellular response to oxidative stress. J Biol Chem 2000;275:7470–7473.
26 Liu H, Lu ZG, Miki Y, Yoshida K: Protein kinase C delta induces transcription of the TP53 tumor suppressor gene by controlling death-promoting factor Btf in the apoptotic response to DNA damage. Mol Cell Biol 2007;27:8480–8491.
27 Voss OH, Kim S, Wewers MD, Doseff AI: Regulation of monocyte apoptosis by the protein kinase Cdelta-dependent phosphorylation of caspase-3. J Biol Chem 2005;280:17371–17379.

28 Soto JL, Castrillo JL, Dominguez F, Dieguez C: Regulation of the pituitary-specific transcription factor GHF-1/Pit-1 messenger ribonucleic acid levels by growth hormone-secretagogues in rat anterior pituitary cells in monolayer culture. Endocrinology 1995;136:3863–3870.

29 Garcia A, Alvarez CV, Smith RG, Dieguez C: Regulation of Pit-1 expression by ghrelin and GHRP-6 through the GH secretagogue receptor. Mol Endocrinol 2001;15:1484–1495.

30 Lin FT, MacDougald OA, Diehl AM, Lane MD: A 30-kDa alternative translation product of the CCAAT/enhancer binding protein alpha message: transcriptional activator lacking antimitotic activity. Proc Natl Acad Sci USA 1993;90:9606–9610.

31 Lin FT, Lane MD: CCAAT/enhancer binding protein alpha is sufficient to initiate the 3T3-L1 adipocyte differentiation program. Proc Natl Acad Sci USA 1994;91:8757–8761.

32 Wang ND, Finegold MJ, Bradley A, et al: Impaired energy homeostasis in C/EBP alpha knockout mice. Science 1995;269:1108–1112.

33 Flodby P, Barlow C, Kylefjord H, Ahrlund-Richter L, Xanthopoulos KG: Increased hepatic cell proliferation and lung abnormalities in mice deficient in CCAAT/enhancer binding protein alpha. J Biol Chem 1996;271:24753–24760.

34 Schaufele F: CCAAT/enhancer-binding protein alpha activation of the rat growth hormone promoter in pituitary progenitor GHFT1-5 cells. J Biol Chem 1996;271:21484–21489.

35 Jacob KK, Stanley FM: CCAAT/enhancer-binding protein alpha is a physiological regulator of prolactin gene expression. Endocrinology 1999;140:4542–4550.

36 Day RN, Voss TC, Enwright JF 3rd, Booker CF, Periasamy A, Schaufele F: Imaging the localized protein interactions between Pit-1 and the CCAAT/enhancer binding protein alpha in the living pituitary cell nucleus. Mol Endocrinol 2003;17:333–345.

37 Enwright JF 3rd, Kawecki-Crook MA, Voss TC, Schaufele F, Day RN: A PIT-1 homeodomain mutant blocks the intranuclear recruitment of the CCAAT/enhancer binding protein alpha required for prolactin gene transcription. Mol Endocrinol 2003;17:209–222.

38 Schuchardt A, D'Agati V, Larsson-Blomberg L, Costantini F, Pachnis V: Defects in the kidney and enteric nervous system of mice lacking the tyrosine kinase receptor Ret. Nature 1994;367:319–320.

39 Kashuk CS, Stone EA, Grice EA, et al: Phenotype-genotype correlation in Hirschsprung disease is illuminated by comparative analysis of the RET protein sequence. Proc Natl Acad Sci USA 2005;102:8949–8954.

40 Fisher S, Grice EA, Vinton RM, Bessling SL, McCallion AS: Conservation of RET regulatory function from human to zebrafish without sequence similarity. Science 2006;312:276–279.

41 Garcia-Lavandeira M, Quereda V, Flores I, et al: A GRFa2/Prop1/stem (GPS) cell niche in the pituitary. PLoS ONE 2009;4:e4815.

42 Quereda V, Malumbres M: Cell cycle control of pituitary development and disease. J Mol Endocrinol 2009;42:75–86.

43 Bilodeau S, Roussel-Gervais A, Drouin J: Distinct developmental roles of cell cycle inhibitors p57Kip2 and p27Kip1 distinguish pituitary progenitor cell cycle exit from cell cycle reentry of differentiated cells. Mol Cell Biol 2009;29:1895–1908.

Clara V. Alvarez
Department of Physiology, School of Medicine, IDIS University of Santiago de Compostela (USC)
c/ San Francisco s/n
ES–15782 Santiago de Compostela (Spain)
Tel. +34 981 582658, Fax +34 981 574145, E-Mail clara.alvarez@usc.es

Testing Growth Hormone Deficiency in Adults

Enrico Gabellieri[a] · Luca Chiovato[a] · Mary Lage[b] · Ana I. Castro[b] · Felipe F. Casanueva[b]

[a]UO of Internal Medicine and Endocrinology, S. Maugeri IRCCS Foundation, Chair of Endocrinology, University of Pavia, Pavia, Italy; [b]Department of Medicine, Santiago de Compostela University, Complejo Hospitalario Universitario de Santiago; CIBER de Fisiopatología Obesidad y Nutrición, Instituto Salud Carlos III, Santiago de Compostela, Spain

Abstract

Growth hormone deficiency (GHD) in adults is a recognized syndrome which is defined biochemically within an appropriate clinical context. Clinically, patients investigated for GHD should include those with signs and symptoms of hypothalamic-pituitary disease, those who have received cranial irradiation or tumor treatment and those with traumatic brain injury or subarachnoid hemorrhage. Patients with three or more pituitary hormone deficiencies and an IGF-I below the reference range do not require provocative testing. The other patients need a provocative test of GH secretory reserve for the diagnosis of GHD. Insulin tolerance test is considered the diagnostic test of choice, however, the GH-releasing hormone (GHRH) + arginine, the GHRH + growth hormone-releasing peptide and the glucagon stimulation tests are well validated alternative tests in adults. Cutoffs differ across tests and results may be influenced by gender, age, body mass index, and the assay reference preparation.

Copyright © 2010 S. Karger AG, Basel

Growth hormone deficiency (GHD) in adults causes abnormalities in lipoprotein and carbohydrate metabolism and in body composition, reduced physical performance, impaired psychological well being, subnormal bone density [1], increased cardiovascular morbidity [2] and premature mortality [3]. GHD in adults is a recognized syndrome in which the replacement therapy with GH improves the impaired health [4, 5]. However, in adults the signs and symptoms of growth hormone deficiency are less clear than in children, and none of them is pathognomonic because many of the signs and symptoms are common to other conditions including normal ageing.

The measurement of GH serum levels and of insulin-like growth factor I (IGF-I), the major peripheral hormone of the somatotroph axis, is not diagnostic for GHD. This is due to the fact that GH is secreted in a pulsatile manner and that a normal decline in GH levels occurs with ageing or with increased adiposity [6]. The serum levels of IGF-I also undergo an age-related decline [6, 7] and are lower in conditions such as malnutrition, hepatic disease, poorly controlled diabetes mellitus or hypothyroidism. Because serum levels of GH and IGF-I of patient with adult-onset GHD can overlap with those of healthy individuals, the diagnosis of GHD must be biochemically established by a provocative test of GH secretory reserve. The concept underlying the use of provocative tests is to compare the peak level of GH response to a stimulatory agent in normal subjects as compared to GHD patients; the final goal being the definition of a cut-off point which may clearly divide the two groups. In the past decades, various stimuli, alone or in combination, have been proposed for GH stimulation tests. Although results of many provocative tests have been published, very few tests have been extensively and rigorously validated.

In order to make diagnostic and treatment criteria more homogeneous, in 1997 the GH Research Society (GRS) convened an international workshop that formulated Consensus Guideline for the Diagnosis and Treatment of GHD in adults, which was adopted internationally [8]. In 2007, a second international workshop has been convened by GRS [9].

The workshops define patients who should be evaluated for GH deficiency in adulthood:
- patients with signs and symptoms of hypothalamic-pituitary disease (endocrine, structural, and/or genetic causes);
- patients who have received cranial irradiation or tumour treatment;
- patients with traumatic brain injury (TBI) or subarachnoid hemorrhage.

Childhood-onset GHD may persist in adulthood, but retesting is needed to confirm the diagnosis. Adult-onset isolated GHD is more common than previously thought [10], particularly among those who have suffered a TBI [11]. However, in patients with traumatic brain injury, since GH secretion may recover, GHD should be tested at least 1 year after the injury. Furthermore, the severity of the TBI does not correlate with the degree of pituitary dysfunction [12].

The workshops underline that severe GHD must be defined biochemically within an appropriate clinical context, in particular the diagnosis of this disorder is made by finding subnormal serum GH response to provocative tests.

One stimulation test is sufficient for the diagnosis of adult GHD. Testing must be performed under stable and adequate replacement for other hormonal deficits in patients with hypopituitarism. Diagnostic testing should be performed in endocrine facilities with considerable experience in such procedures. Not all patients suspected of having GHD, however, require a GH stimulation test for diagnosis. Indeed, GHD is certain in patients with three or more pituitary hormone deficiencies and an IGF-I below the reference range, and no conditions that would otherwise lower IGF-I [13–15] (fig. 1). For the diagnosis of adult GHD, the measurement of other markers of GH secretion does not offer any significant advantage compared to the determination of total IGF-I levels [13].

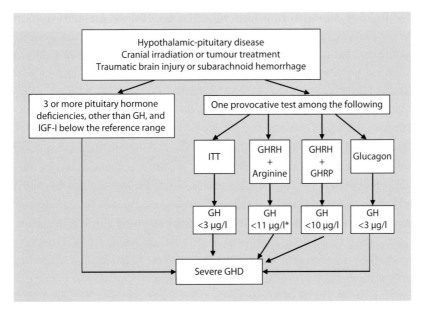

Fig. 1. Diagnosis of severe GHD in adults. *Related to BMI <25.

The 1997 workshop recommended the insulin tolerance test (ITT) as the diagnostic test of choice. However, other tests are as reliable as ITT in the diagnosis of GHD. In particular, the 2007 workshop agreed that GHRH + arginine, GHRH + growth hormone-releasing peptide (GHRP) and glucagon stimulation tests are also well validated tests in adults. Each test has advantages and limitations [16] and each shows intraindividual variability.

The ITT evaluates the integrity of the hypothalamic-pituitary axis, provided adequate hypoglycemia is achieved. This test is contraindicated in patients with electrocardiografic evidence or history of ischemic heart disease or in patients with seizure disorders, brain injured patients and in the elderly [17]. Given these precautions, the test is safe but is the most unpleasant for the patient. For ITT the validated cut-off for GHD in adults is a peak GH response <3 μg/l. ITT shows the greatest sensitivity and specificity within the first 5 years after irradiation, a frequent cause of hypothalamic GHD, and therefore is the provocative test of choice in these patients. Indeed, because the combined tests stimulate both the hypothalamus and the pituitary, GHD due to hypothalamic disease may be missed.

Although ITT is considered the gold standard test for GHD in adults, some concerns about its reproducibility and specificity have been reported [18–20]. Moreover, results of ITT are gender-dependent and strongly influenced by age and adiposity [21].

The glucagon test has been confirmed as a reliable diagnostic test and the cut-off for GHD in adults is a peak GH response <3 μg/l [22, 23]. As for ITT, results of glucagons test are also gender-dependent and strongly influenced by age and adiposity.

The GHRH + arginine stimulation a well validated test, which presents a good safety profile, being caracterized by the lack of contraindications, with the notable exception of chronic renal failure. Results of the GHRH + arginine test are not altered by age, but are affected by adiposity. Therefore, taking into account BMI, the following cut-off levels for GHD in adults have been validated for the GHRH + arginine test: for a BMI <25 a peak of GH <11 µg/l; for a BMI 25–30 a peak of GH <8 µg/l, for a BMI >30 a peak of GH <4 µg/l.

Similarly to the GHRH + arginine test, the GHRH + GHRP test displays a good safety profile and no contraindication [24, 25], but its results are also affected by adiposity only at BMI higher than 35 [26]. For this test has been proposed a cut-off point ≤10 GH µg/l for the diagnosis of GHD.

Other tests such as the clonidine, the L-DOPA, and the arginine alone stimulation are considered not useful in adults. Ghrelin itself, the natural ligand of GH secretagogues receptor, is likely to represent a good provocative test [27]. Ghrelin mimetics alone are currently under evaluation as a test of the GH axis.

It is important to underline that some limitations exist in the GH and IGF-I assays. The GH cut-off value that distinguishes GH sufficiency from GH deficiency is defined employing radioimmunoassays (RIAs). However, the results obtained vary between different assay methods, and therefore the cut-off value may need to be adjusted appropriately. This problem became more acute after 1997 when the polyclonal radioimmunoassay were replaced by highly sensitive 2-site monoclonal assays, which normally detect lower amounts of GH (≤50%) than the polyclonal assays.

The problem of standardization of immunoassays involves a combination of problems [28]. GH is not a homogenous molecule indeed, GH in circulation consists of a wide variety of molecular isoforms [29, 30]. Monoclonal antibodies used in immunoassay recognize specific epitopes present on the surface of the antigen. This can explain the discrepancy between assays results using polyclonal or monoclonal antibodies. The high affinity of GH to GH-binding protein (GHBP), which circulates in serum [31], could lead to underestimating GH concentrations because some epitopes might not be accessible for certain antibodies. Standard preparations used to calibrate the assay changed through the time and this change has had a major influence on the absolute concentrations reported by different assays. Moreover, need to convert between the different units used in GH assay provides an additional source of problems in comparing GH assays results. The use of a single, recombinant calibrator for all GH assays (International Reference preparation (IRP) 98/574) and the reporting of all GH assay results in mass units (µg/l) of the recombinant calibrator are suggested as a first step toward a standardization [32, 33].

Similar problems concern the measurement of IGF-I. An accurate measure of IGF-I and an universal calibrator are essential. The latter should be a preparation of highly purified recombinant human IGF-I. The utility of IGF-I measurements will be enhanced by specifying age and gender normal ranges.

Conclusions

Testing for GHD should be undertaken with the intention to treat patients with an appropriate clinical context. Insulin tolerance test, combined administration of GHRH with arginine or GHRP, and glucagon are validated tests for the diagnosis of GHD in adults. Low IGF-I is a reliable diagnostic indicator of GHD, however a normal IGF-I does not rule out GHD. Universally adopted calibrators for GH and IGF-I assays are required.

References

1 Gómez JM, Gómez N, Fiter J, Soler J: Effects of long-term treatment with GH in the bone mineral density of adults with hypopituitarism and GH deficiency and after discontinuation of GH replacement. Horm Metab Res 2000;32:66–70.
2 Colao A, Di Somma C, Cuocolo A, Filippella M, Rota F, Acampa W, Savastano S, Salvatore M, Lombardi G: The severity of growth hormone deficiency correlates with the severity of cardiac impairment in 100 adult patients with hypopituitarism: an observational, case-control study. J Clin Endocrinol Metab 2004;89:5998–6004.
3 Carroll PV, Christ ER, Bengtsson BA, Carlsson L, Christiansen JS, Clemmons D, Hintz R, Ho K, Laron Z, Sizonenko P, Sönksen PH, Tanaka T, Thorne M: Growth hormone deficiency in adulthood and the effects of growth hormone replacement: a review. Growth Hormone Research Society Scientific Committee. J Clin Endocrinol Metab 1998;83:382–395.
4 Verhelst J, Abs R: Long-term growth hormone replacement therapy in hypopituitary adults. Drugs 2002;62:2399–2412.
5 Amato G, Carella C, Fazio S, La Montagna G, Cittadini A, Sabatini D, Marciano-Mone C, Saccá L, Bellastella A: Body composition, bone metabolism, and heart structure and function in growth hormone (GH)-deficient adults before and after GH replacement therapy at low doses. J Clin Endocrinol Metab 1993;77:1671–1676.
6 Rudman D, Kutner MH, Rogers CM, Lubin MF, Fleming GA, Bain RP: Impaired growth hormone secretion in the adult population: relation to age and adiposity. J Clin Invest 1981;67:1361–1369.
7 Clemmons DR: Commercial assays available for insulin-like growth factor I and their use in diagnosing growth hormone deficiency. Horm Res. 2001;55(suppl 2):73–79.
8 Consensus Guidelines for the Diagnosis and Treatment of Adults with Growth Hormone Deficiency: Summary statement of the Growth Hormone Research Society Workshop on Adult Growth Hormone Deficiency. J Clin Endocrinol Metab 1998;83:379–381.
9 Ho KK, 2007 GH Deficiency Consensus Workshop Participants: Consensus guidelines for the diagnosis and treatment of adults with GH deficiency II: a statement of the GH Research Society in association with the European Society for Pediatric Endocrinology, Lawson Wilkins Society, European Society of Endocrinology, Japan Endocrine Society, and Endocrine Society of Australia. Eur J Endocrinol 2007;157:695–700.
10 Abs R, Mattsson AF, Bengtsson BA, Feldt-Rasmussen U, Góth MI, Koltowska-Häggström M, Monson JP, Verhelst J, Wilton P, KIMS Study Group: Isolated growth hormone (GH) deficiency in adult patients: baseline clinical characteristics and responses to GH replacement in comparison with hypopituitary patients: a sub-analysis of the KIMS database. Growth Horm IGF Res 2005;15:349–359.
11 Tanriverdi F, Unluhizarci K, Kocyigit I, Tuna IS, Karaca Z, Durak AC, Selcuklu A, Casanueva FF, Kelestimur F: Brief communication: pituitary volume and function in competing and retired male boxers. Ann Intern Med. 2008;148:827–831.
12 Agha A, Rogers B, Sherlock M, O'Kelly P, Tormey W, Phillips J, Thompson CJ: Anterior pituitary dysfunction in survivors of traumatic brain injury. J Clin Endocrinol Metab 2004;89:4929–4936.
13 Aimaretti G, Corneli G, Razzore P, Bellone S, Baffoni C, Bellone J, Camanni F, Ghigo E: Usefulness of IGF-I assay for the diagnosis of GH deficiency in adults. J Endocrinol Invest 1998;21:506–511.

14 Aimaretti G, Corneli G, Baldelli R, Di Somma C, Gasco V, Durante C, Ausiello L, Rovere S, Grottoli S, Tamburrano G, Ghigo E: Diagnostic reliability of a single IGF-I measurement in 237 adults with total anterior hypopituitarism and severe GH deficiency. Clin Endocrinol (Oxf) 2003;59:56–61.
15 Hartman ML, Crowe BJ, Biller BM, Ho KK, Clemmons DR, Chipman JJ; HyposCCS Advisory Board, US HypoCCS Study Group: Which patients do not require a GH stimulation test for the diagnosis of adult GH deficiency? J Clin Endocrinol Metab 2002;87:477–485.
16 Biller BM, Samuels MH, Zagar A, Cook DM, Arafah BM, Bonert V, Stavrou S, Kleinberg DL, Chipman JJ, Hartman ML: Sensitivity and specificity of six tests for the diagnosis of adult GH deficiency. J Clin Endocrinol Metab 2002;87:2067–2079.
17 Ghigo E, Aimaretti G, Corneli G: Diagnosis of adult GH deficiency. Growth Horm IGF Res 2008;18:1–16.
18 Jones SL, Trainer PJ, Perry L, Wass JA, Besssser GM, Grossman A: An audit of the insulin tolerance test in adult subjects in an acute investigation unit over one year. Clin Endocrinol (Oxf) 1994;41:123–128.
19 Hoeck HC, Vestergaard P, Jakobsen PE, Laurberg P: Test of growth hormone secretion in adults: poor reproducibility of the insulin tolerance test. Eur J Endocrinol 1995;133:305–312.
20 Vestergaard P, Hoeck HC, Jakobsen PE, Laurberg P: Reproducibility of growth hormone and cortisol responses to the insulin tolerance test and the short ACTH test in normal adults. Horm Metab Res 1997;29:106–110.
21 Casanueva FF, Pombo M, Leal A, Popovic V, Dieguez C: Biochemical diagnosis of growth hormone deficiency in adults; in Abs R, Feldt-Rasmussen U (eds): Growth Hormone Deficiency in Adults: 10 Years of KIMS. Oxford, Oxford PharmaGenesis Ltd, 2004, pp 91–101.
22 Gómez JM, Espadero RM, Escobar-Jiménez F, Hawkins F, Picó A, Herrera-Pombo JL, Vilardell E, Durán A, Mesa J, Faure E, Sanmartí A: Growth hormone release after glucagon as a reliable test of growth hormone assessment in adults. Clin Endocrinol (Oxf) 2002;56:329–334.
23 Conceição FL, da Costa e Silva A, Leal Costa AJ, Vaisman M: Glucagon stimulation test for the diagnosis of GH deficiency in adults. J Endocrinol Invest 2003;26:1065–1070.
24 Popovic V, Leal A, Micic D, Koppeschaar HP, Torres E, Paramo C, Obradovic S, Dieguez C, Casanueva FF: GH-releasing hormone and GH-releasing peptide-6 for diagnostic testing in GH-deficient adults. Lancet 2000;356:1137–1142.
25 Popovic V, Pekic S, Doknic M, Micic D, Damjanovic S, Zarkovic M, Aimaretti G, Corneli G, Ghigo E, Deiguez C, Casanueva FF: The effectiveness of arginine + GHRH test compared with GHRH + GHRP-6 test in diagnosing growth hormone deficiency in adults. Clin Endocrinol (Oxf) 2003;59:251–257.
26 Kelestimur F, Popovic V, Leal A, Van Dam PS, Torres E, Perez Mendez LF, Greenman Y, Koppeschaar HP, Dieguez C, Casanueva FF: Effect of obesity and morbid obesity on the growth hormone (GH) secretion elicited by the combined GHRH + GHRP-6 test. Clin Endocrinol (Oxf) 2006;64:667–671.
27 Aimaretti G, Baffoni C, Broglio F, Janssen JA, Corneli G, Deghenghi R, van der Lely AJ, Ghigo E, Arvat E: Endocrine responses to ghrelin in adult patients with isolated childhood-onset growth hormone deficiency. Clin Endocrinol (Oxf) 2002;56:765–771.
28 Bidlingmaier M: Problems with GH assays and strategies toward standardization. Eur J Endocrinol. 2008;159(suppl 1):S41–S44.
29 Baumann G: Growth hormone heterogeneity: genes, isohormones, variants, and binding proteins. Endocr Rev 1991;12:424–449.
30 Baumann G: Growth hormone heterogeneity in human pituitary and plasma. Horm Res 1999;51(suppl 1):2–6.
31 Baumann G: Growth hormone binding protein 2001. J Pediatr Endocrinol Metab 2001;14:355–375.
32 Trainer PJ, Barth J, Sturgeon C, Wieringaon G: Consensus statement on the standardization of GH assays. Eur J Endocrinol 2006;155:1–2.
33 Sheppard MC: Growth hormone assay standardization: an important clinical advance. Clin Endocrinol (Oxf) 2007;66:157–161.

Felipe F. Casanueva
School of Medicine, Santiago de Compostela University
Calle San Francisco SN, PO Box 563
ES–15780 Santiago de Compostela (Spain)
Tel. +34 981 572121, Fax +34 919 966 6025, E-Mail endocrine@usc.es

Serum Insulin-Like Growth Factor-1 Measurement in the Diagnosis and Follow-up of Patients with Acromegaly: Preliminary Data

Mirtha Guitelman[a,b] · Graciela Radczuk[c] ·
Natalia García Basavilbaso[a,b] · Adriana Oneto[b,c] · Armando Basso[b]

[a]División Endocrinología Hospital Carlos G. Durand, [b]Instituto de Neurocirugía de Buenos Aires (INBA), and [c]Laboratorio TCba Salguero, Buenos Aires, Argentina

Abstract

Measurement of serum insulin-like growth factor-1 (IGF-I) is the current method for diagnosing and monitoring acromegaly. However, the use of commercially available kits needs to be validated. In our study, we have investigated the use of two different IGF-I immunoassays in patients already diagnosed with acromegaly. We compared a two-site immunoradiometric assay with ethanol-acid extraction (IRMA-DSL) and a solid-phase chemiluminescent immunometric assay (ICMA-IMMULITE), correlating the clinical finding with the biochemical results. A total of 102 samples (77 women and 25 men aged 18–79 years) were analyzed with the two different IGF-I assays. Sixty-four of samples had been taken from patients with acromegaly in different stages. Pearson regression showed a high correlation coefficient; otherwise, Bland and Altman analyses showed a mean difference of 177.6 ng/ml, with upper and lower limits of –183.5 and 538.7 ng/ml in the 102 samples studied. Normal serum IGF-I was found in 64 and 41.5% of patients with treated acromegaly when measured by ICMA and IRMA, respectively. In our study, IGF-I-ICMA had a better clinical correlation in patients with treated acromegaly. The reevaluation of current IGF-I immunoassays is necessary to correctly interpret treatment response in acromegalic patients and thus achieve a better correlation between clinical and biochemical results.

Copyright © 2010 S. Karger AG, Basel

Acromegaly is the condition that results from prolonged and excessive circulating growth hormone (GH) levels in adults. GH primarily assists in the synthesis of peripheral insulin-like growth factor-I (IGF-I), mostly in hepatocytes, a process that leads to cell proliferation and inhibition of apoptosis.

IGF-I is a useful tool for the diagnosis, postoperative assessment of regression or 'cure', and long-term monitoring of active acromegaly [1, 2]. The increased rates of

Table 1. Serum IGF-1 assays: characteristics

IGF-1 Assay	Signal	Antibodies	Reference preparation (WHO IRP)	Extraction method
ICMA DPC-IMM	chemiluminescent	monoclonal/polyclonal	87/518	no
IRMA DSL	isotopic	monoclonal/polyclonal	87/518	acid/alcohol

morbidity and mortality associated with acromegaly make reliable methods of assessment essential [3].

GH acts mostly through IGF-I. This is supported by the fact that, in patients with GH insensitivity, exogenous IGF-I promotes growth and may even mimic the phenotype of acromegaly, indicating that it is IGF-I and not GH the true growth hormone in this disease [4]. In addition, the use of the GHR antagonist Pegvisomant, which blocks signal transduction and transcription of serum IGF-I, results in resolution of symptoms, signs, and metabolic features of acromegaly, despite a persistently elevated GH [5, 6].

There are some physiological factors that change IGF-I levels, mainly age and binding proteins [7]. The IGF-I surge in puberty results from values outside 'normal' age-related reference ranges. IGF-I also shows an age-dependent decrease of 10–16% per decade, even in acromegaly [8]. Ninety-nine percent of the circulating IGF-I is bound together with IGFBP-3 and the acid-labile subunit (ALS) in ternary, 150-kDa complexes. This complexation prolongs the half-life (15 h) of IGF-I and prevents its transfer out of the vascular space [9, 10].

There are several problems associated with serum IGF-I measurements, including difficulty in comparing results among laboratories due to a lack of standardization, susceptibility to interference from IGF-binding proteins (IGFBP), lack of a pure international reference preparation, and the need for appropriate age-adjusted normative data. Current commercially available immunoassays differ in terms of assay principle (competitive vs. noncompetitive), assay format (manual vs. automated), type of antibodies used (monoclonal vs. polyclonal) and label used (radioactive, enzyme-linked detection or chemiluminescent detection) [11, 12].

In addition, assays can be divided, according to the method used to avoid interference from IGFBP, into nonextraction and extraction methods [13]. Such pitfalls in the fine-tuning of this hormone, together with the clinical discrepancies we found, led us to investigate two different immunoassays in patients with GH excess.

The most widely used immunoassay in our country until 2007 was the two-site immunoradiometric assay with ethanol-acid extraction, IRMA-DSL. With regard to patient follow-up during treatment, we found discrepancies between the clinical score and biochemical results. Then we decided to include a new immunoassay in order to establish the cause of those discrepancies.

Objective

In our study, we investigated and compared two different commercially available IGF-I immunoassays in patients with known acromegaly, and correlated their clinical finding and biochemical results. We also included a small group of normal subject as control.

Patients and Methods

Subjects and Patients

A total of 102 samples (77 women and 25 men, 18–79 years old) were analyzed using two different IGF-I assays. Sixty-four of the samples had been taken from acromegalic patients, 44 of which were under different medical treatments (35 were receiving somatostatin analogs, 5 pegvisomant, and 4 cabergoline), 11 had active disease and were under no medical treatment, and 9 were patients who had attained postoperative cure (1 had undergone radiation therapy). Four additional samples had been taken from 4 subjects diagnosed with acromegaly due to high levels of IGF-I measured by IRMA; all of them were asymptomatic, showed a normal pituitary MRI, and had GH levels <1 ng/ml in the oral glucose tolerance test (OGTT).

Acromegaly was diagnosed on account of the findings of both lack of GH suppression during OGTT (GH >1 µg/l) and higher than normal age-matched serum IGF-I levels, along with clinical features of GH excess and pathological MRI findings [14]. Acromegaly is considered to be controlled or cured based on biochemical parameters, including GH levels lower than 1 ng/ml in the OGTT and normal age-matched IGF-I [14–15].

Our control group consisted of 38 samples from nonacromegalic patients.

Laboratory Methods

Assay 1. Two-site immunoradiometric assay with ethanol-acid extraction, IRMA-DSL.
Assay 2. Solid-phase chemiluminescent immunometric assay, ICMA-IMMULITE.
Reference ranges for each method were used as provided by the manufacturers. Immunoassays characteristics are shown in table 1.
Statistical analyses were performed by Pearson correlation and Bland-Altman tests.

Results

Pearson regression showed a correlation coefficient of r = 0.8165 (significance level p< 0.0001; fig. 1). Bland and Altman analyses showed a mean difference of 177.6 ng/ml, with upper and lower limits of –183.5 and 538.7 ng/ml in the 102 samples studied (table 2).

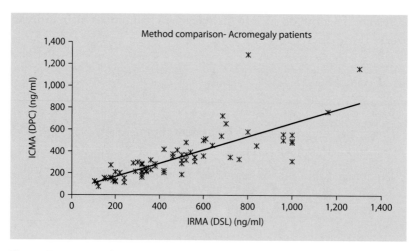

Fig. 1. Pearson correlation between ICMA and IRMA of all samples analyzed.

Normal serum IGF-I was found in 64% and 41.5% of patients with treated acromegaly (surgery, medical treatment or both) when measured by ICMA and IRMA, respectively. Four patients, considered as cured, with GH levels <1 ng/ml in the OGTT, normal MRI findings, and no symptoms or signs of active disease, showed high IRMA, but normal ICMA IGF-I levels.

When patients receiving somatostatin analogs were analyzed separately, IGF-I was found to be within the normal ranges in 58% (ICMA) and 32% (IRMA).

Improvement of the clinical score was seen in 79% of all treated patients and in 77% of patients under octreotide. Four subjects who had high levels of IRMA-IGF-I but no other suspicious findings of acromegaly showed a normal ICMA-IGF-I, in accordance with their clinical status and the lack of other markers of acromegaly (GH levels <1 µg/l in the OGTT and normal MRI findings).

The 4 additional patients suspected to have acromegaly had normal ICMA IGF-I.

Discussion

Measurement of circulating total IGF-I levels provides an important tool for diagnosing GH disorders as well as for monitoring treatment efficacy; however, several problems can diminish its clinical usefulness. Serum IGF-I levels vary depending on biological determinants and assay performance. This variability may have its physiological source in features such as nutritional status, pubertal stage, pregnancy, age, and gender, all of which should be taken into account when interpreting IGF-I values [16, 17].

Our findings suggest that the IGF-I-ICMA had a better clinical correlation than the IRMA method in patients with treated acromegaly. IGF-I-IRMA showed a low rate

Table 2. Bland-Altman analysis of 102 samples from patients and normal subjects

	Pearson correlation coefficient	Ranges ng/ml (IRMA)	Ranges ng/ml (ICMA)	Bland and Altman analyses		
				median	lower limit	upper limit
All groups (n = 102)	0.8165 (p < 0.0001)	40–1,300	34–1,285	177.6	–183.5	538.7
Acromegaly group (n = 64)	0.7879 (p < 0.0001)	106–1,300	75–1,285	136	–190.7	462.7
Control group (n = 38)	0.8457 (p < 0.0001)	40–1,000	34–482	89.1	–119.5	297.8

Difference between IGF-I concentration determined by IRMA and ICMA assay.

of IGF-I normalization in patients with treated acromegaly; conversely, the results obtained with ICMA are in accordance with previously published data [18].

A potential pitfall of IRMA resides in its false-positive results, which make it a less-reliable marker for screening GH excess.

Prokajac et al. [19] reported that two serum samples known to have shown borderline results were sent to all centers participating in the UK National External Quality Assessment Service (NEQAS). Sample B, with a clinical scenario (i.e. a patient with slight symptoms of acromegaly), was sent to 23 centers measuring IGF-I by means of 6 different immunoassays. There was a 50% variation in the upper limit of the reference ranges between centers, so that 30% of the IGF-I results ruled out the diagnosis.

Furthermore, we found a significant intermethod bias in the upper limit of the reference ranges but less variation in the lower limit. A poor performance of the SDL standard curve at high values could explain this discrepancy.

Granada et al. [20] compared four different immunoassays in patients with acromegaly and found a high variability in IGF-I normative data according to the assay used, as previously reported in the literature. Higher values were reported with the nonextraction IRMAs. In our study, all acromegalic patients had IGF-I >2 SD scores regardless of the assay used or the source of the reference population. Therefore, these results could not be extrapolated to assess disease activity in other patient groups with partially cured acromegaly. In fact, Masart et al. [21] studied the performances of four immunoassays in the follow-up of 40 treated acromegaly patients and found significant differences in the prevalence of normal IGF-I levels according to the assay and the source of the reference values used.

In spite of a high correlation between these two IGF-I methods, the differences found in Bland and Altman analyses suggest that exist analytical variables, namely the number of normal subjects needed to establish a normal range, binding proteins interference, standard and antibodies used for each immunoassay, which should all be taken into account to achieve a correct interpretation of the results.

One of the drawbacks of our study was the small number of subjects in the control group and the lack of validation of each immunoassay reference ranges with an adequate number of normal subjects from our own population. Nevertheless, this small research showed findings similar to those found by others in terms of methodological pitfalls regarding the measurement of serum IGF-I.

The reevaluation of current IGF-I immunoassays is necessary to correctly interpret treatment response in acromegalic patients and thus achieve a better correlation between clinical and biochemical results.

In our study, the solid-phase chemiluminescent immunometric assay ICMA-IMMULITE was superior to the immunoradiometric assay with ethanol-acid extraction IRMA-DSL for diagnosing and monitoring acromegaly.

Currently, we are using ICMA-IMMULITE, which has been validated with almost 1,000 normal subjects. Thus, we have established a reliable local normative for this methodology [data not publ.].

References

1 Brooke AM, Drake WM: Serum IGF-1 levels in the diagnosis and monitoring of acromegaly. Pituitary 2007;10:173–179.
2 Melmed S, Casanueva FF, Cavagnini F, Chanson P, Frohman L, Grossman A, Ho K, Kleinberg D, Lamberts S, Laws E, Lombardi G, Vance ML, Werder KV, Wass J, Giustina A: Acromegaly Treatment Consensus Workshop Participants: Guidelines for acromegaly management. J Clin Endocrinol Metab 2002;87:4054–4058.
3 Holdaway I, Rajasoorya C, Gamble G, Stewart A: Long term treatment outcome in acromegaly. J Clin Endocrinol Metab 2004;89:2789–2796.
4 Walker JL, Crock PA, Behncken SN, et al: A novel mutation affecting the interdomain link region of the growth hormone receptor in a Vietnamese girl, and response to long-term treatment with recombinant human insulin-like growth factor-I and luteinizing hormone releasing hormone analogue. J Clin Endocrinol Metab 1998;83:2554–2561.
5 Trainer PJ, Drake WM, Katznelson L, et al: Treatment of acromegaly with the growth hormone-receptor antagonist pegvisomant. N Engl J Med 2000;342:1171–1177.
6 Van der Lely A, Hutson R, Trainer P, et al: Long term treatment of acromegaly with pegvisomont, a growth hormone receptor antagonist. Lancet 2001;358:1754–1759.
7 Clemmons DR: IGF-I assays: current assay methodologies and their limitations. Pituitary 2007;10:121–128.
8 Elmlinger MW, Kuhnel W, Weber MM, Ranke MB: Reference ranges for two automated chemiluminescent assays for serum insulin-like growth factor I (IGF-I) and IGF binding protein 3 (IGFBP-3) Clin Chem Lab Med 2004;42:654–664.
9 Ranke MB, Emlinger M: Functional role of insulin-like growth factor binding proteins. Hormone Research, 1997;48(suppl 4):9–15.
10 Daughaday WH, Kapadia M, Mariz I: Serum somatomedin binding proteins: physiologic significance and interference in radioligand assay. J Lab Clin Med 1987;109:355–363.
11 Ranke MB, Feldt-Rasmussen U, Bang P, Baxter RC, Camacho- Hubner C, Clemmons DR, Juul A, Orskov H, Strasburger CJ: How should insulin-like growth factor I be measured? A consensus statement. Horm Res 2001;55(suppl 2):106–109.
12 Brooke M, Drake WM: Serum IGF-I levels in the diagnosis and monitoring of acromegaly. Pituitary 2007;10:173–179.
13 Brabant GH: Wallaschofski Normal levels of serum IGF-I: determinants and validity of current reference ranges. Pituitary 2007;10:129–133.
14 Growth Hormone Research Society and Pituitary Society: Biochemical assessment and long-term monitoring in patients with acromegaly: statement from a joint consensus conference of the Growth Hormone Research Society and the Pituitary Society. J Clin Endocrinol Metab 2004;89:3099–3102.

15 Giustina A, Barkan A, Casanueva FF, et al: Criteria for cure of acromegaly: a consensus statement. J Clin Endocrinol Metab 2000;85:526–529.
16 Thissen JP, Ketelslegers JM, Underwood LE: Nutritional regulation of the insulin-like growth factors. Endocr Rev 1994;15:80–101.
17 Gomez JM, Maravall FJ, Gomez N, Navarro MA, Casamitjana R, Soler J: The IGF-I system component concentrations that decrease with ageing are lower in obesity in relationship to body mass index and body fat. Growth Horm IGF Res 2004;14:91–96.
18 Freda PU: Somatostatin analogs in acromegaly. J Clin Endocrinol Metab 2002;87:3013–3018.
19 Pokrajac A, Wark G, Ellis AR, Wear J, Wieringa GE, Trainer PJ: Variation in GH and IGF-I assays limits the applicability of international consensus criteria to local practice. Clin Endocrinol 2007;67:65–70.
20 Granada ML, Ulied A, Casanueva FF, Pico A, Lucas T, Torres E, Sanmartí A: Serum IGF-I measured by four different immunoassays in patients with adult GH deficiency or acromegaly and in a control population. Clin Endocrinol 2007;Journal Compilation. Oxford, Blackwell, 2007.
21 Massart C, Poirier JY: Serum insulin-like growth factor-I measurement in the follow-up of treated acromegaly: comparison of four immunoassays. Clin Chim Acta 2006;373:176–179.

Dra. Mirtha Guitelman
División Endocrinología Hospital Carlos G. Durand
Díaz Velez 5044
C1405DCS Buenos Aires (Argentina)
Tel. +54 11 4982 5212, Fax +54 11 4958 4377, E-Mail mguitelman@speedy.com.ar

Diagnosis of Cure in Cushing's Syndrome: Lessons from Long-Term Follow-Up

M.J. Barahona · E. Resmini · N. Sucunza · S.M. Webb

Endocrinology and Medicine Departments and Centro de Investigación Biomédica en Red de Enfermedades Raras (CIBER-ER, Unidad 747), ISCIII, and Hospital Sant Pau, Universitat Autònoma de Barcelona, Barcelona, Spain

Abstract

It is generally assumed that endocrine 'cure' of hypercortisolism after successful treatment for Cushing's syndrome (CS) is associated with reversal of increased morbidity and mortality, typical of the active disease. However, recent data do not support this idea; increased cardiovascular risk is still present 5 years after endocrine cure, and health-related quality of life (HRQoL), although improved when compared to the active phase of hypercortisolism, is still impaired when compared to normal population. Abnormal body composition typical of hypercortisolism (i.e., increased total and trunk fat, reduced bone mass and lean body mass) is not completely normalized, even years after controlling hypercortisolism. Thus, control of hypercortisolism in CS does not normalize HRQoL, long-term cardiovascular risk and morbidity, body composition nor some metabolic parameters. Whether the same occurs in patients exposed to pharmacological doses of exogenous glucocorticoids, and whether the body composition abnormalities associated with the exposure to exogenous glucocorticoids are reversible or not, are worth considering.

Copyright © 2010 S. Karger AG, Basel

The prognosis of Cushing's syndrome (CS) has improved over the years; however, patients may be exposed to hypercortisolism for a long time before definite treatment is efficient. The long-term consequences of prior chronic hypercortisolism are currently unclear, although there is some evidence that it does impair health. Historically, both morbidity and mortality were increased in patients who suffered CS [1–4], and the main reported cause of death is cardiovascular. Since it is a rare disease, it is difficult to obtain long series where this issue may be approached, and contributes to explain some discordant results [1]. Others have not observed increased mortality [5, 6]. One study carried out in 15 patients with CS and 2 control groups (sex- and age-matched and BMI-matched controls) of 30 individuals showed that despite attaining normal cortisol, hypertension, low HDL-cholesterol, diabetes mellitus, insulin resistance and carotid atheromatosis symptoms persisted after 5 years [7]. It has recently been demonstrated that health-related quality of life (HRQoL) is severely affected in CS, and is influenced, among other things, by disease duration, gender and age, and may not normalize despite hormonal 'control' of the disease [8, 9]. In this paper we report the morbidity at diagnosis of CS and mortality and further morbidity several years after treatment for hypercortisolism in our cohort of CS; furthermore, in a collaborative study with different European reference centers, we have performed a transversal evaluation of HRQoL in patients who have suffered CS [9].

Retrospective Study of Morbidity and Mortality in Our Cohort of Cushing's Syndrome

We retrospectively reviewed the clinical records of patients with endogenous CS treated at our hospital since 1982. Patients with pituitary-dependent disease, as well as adrenal tumors and ectopic ACTH-secreting tumors were included, but not adrenal carcinomas due to their worse prognosis. A total of 98 CS patients were identified (mean age at diagnosis of CS: 38 ± 13 years, 83 women, 73 of pituitary origin, 23 of adrenal origin – 2 of which were bilateral and 2 of ectopic origin); of these 98 patients, 5 had died and 32 were lost to follow-up (including patients from other regions of Spain who after successful operation were in remission of hypercortisolism and followed by their previous physicians); of the remaining 61 patients, 17 were hypercortisolemic (14 women, 4 of which had recurred; 14 of pituitary and 3 of adrenal origin) and 44 were in remission (37 women, 34 of pituitary and 10 of adrenal origin). The prevalence of cardiovascular risk factors in our cohort of 98 patients at diagnosis of CS is summarized in table 1; 18% of patients had overt diabetes, 63% arterial hypertension and 38% dyslipidemia. For comparison, the same data corresponding to our acromegaly cohort, followed in our center since 1982 are presented. While prevalence of hypertension and dyslipidemia are similar, diabetes was more prevalent in acromegaly than in CS, indicating a greater effect of GH than cortisol on glucose intolerance.

Prospective Case-Control Study of Our Patients with Cushing's Syndrome

We also conducted a case-control study including the 61 patients with CS diagnosed and controlled in our centre (17 active and 44 cured) and 106 healthy gender- and age-matched controls which were selected among the blood donors database at our hospital. Letters were sent to selected controls, and a phone call was made 1 week later; the first control to accept was included. Controls that referred glucocorticoid (GC) treatment or malignant disease were excluded. All patients signed an informed consent after study approval by the hospital ethics committee. In cured CS, mean age at the time of study was 50 ± 14 years, while mean age at the time of CS diagnosis was 36 ± 12 years; mean time of hormonal cure of hypercortisolism was 11 ± 6 years. Mean BMI was 28 ± 6 and waist circumference 94 ± 14 cm. In these patients,

Table 1. Morbidity at diagnosis of CS and mortality and morbidity after follow-up in patients operated and followed since 1982; for comparison the same data corresponding to the acromegaly cohort of our center (Hospital S Pau, Barcelona) are also shown

	CS, % (n = 98)	Acromegaly, % (n = 99)
At diagnosis		
Diabetes mellitus	18	34
Hypertension	63	62
Dyslipidemia	38	33
At follow-up		
Ischemic heart disease	3	4
Stroke	3	2.5
Thrombosis	6	0
Peripheral arterial vascular disease	2	2.4
Death/cardiovascular cause	5/60	10/56
Median age at death	49	62
Mean age at diagnosis, years	38 ± 13	42 ± 12
Mean follow-up, years	11 ± 6	14.5 ± 9.3

whole body composition was evaluated by dual-energy X-ray absorptiometry scanning (DEXA, Delphi QDR 4500, Hologic).

Statistical analysis was performed using SPSS 15.0 statistical package for Windows (SPSS Inc., Chicago, Ill., USA). Quantitative data are expressed as mean and SD (Gaussian distribution) or as median and range (non-Gaussian distribution), and qualitative data, as percentages. Data distribution was analyzed by the Kolmogorov-Smirnov test. Comparisons between 2 groups were performed using Student's t (Gaussian distribution) and Mann-Whitney's U (non-Gaussian distribution) tests. A χ^2 test was performed for categorical variables.

The cardiovascular (CV) morbidity of our cohort of patients after a mean 11 years of follow-up was important (table 1), namely 3% of ischemic heart disease, 3% of stroke, 6% thrombosis and 2% peripheral arterial vascular disease. The mortality of our series was 5% (5 patients). One male with noncured Cushing's disease (CD) died of a ruptured aortic aneurism at age 36, a cured female with CD aged 38 from mesenteric thrombosis, and another cured female with CD suffered a sudden cardiac death at age 55. The other 2 patients (one noncured female with ectopic ACTH secretion and one cured female with bilateral macronodular hyperplasia) died at age 56 from progression of tumoral disease, and at age 61 from sepsis after acute pancreatitis,

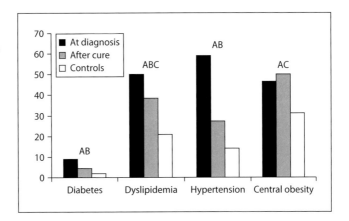

Fig. 1. Prevalence of cardiovascular risk factors in patients with CS at diagnosis and after long-term cure and comparison with that of healthy age- and sex-matched controls. A = p < 0.05 between patients with CS at diagnosis and controls; B = p < 0.05 between patients with CS at diagnosis and after cure; C = p < 0.05 between patients after cure of CS and controls.

respectively. Thus, median age at death was 49 years, 3 of the 5 deaths were of CV origin (of which 2 were cured) and the 3 patients with cured CS had normal cortisol for a mean of 9.5 years at the time of death. It would appear therefore, that serious cardiovascular damage persisted, years after control of hypercortisolism.

The comparison between CS and acromegaly regarding mortality and cardiovascular morbidity is also represented in table 1. In the case-control study group (n = 61), we analyzed the prevalence of CV risk factors only in cured patients (before and after cure) and compared it with gender- and age-matched controls. The prevalence of diabetes, dyslipidemia and hypertension was higher at diagnosis of CS than after cure and in controls (9 vs. 4.5 and 2%, p < 0.05; 50 vs. 38.6 and 21%, p < 0.001; 59 vs. 27.3 and 14%, p < 0.01). Controls had less prevalence of central obesity than patients at diagnosis of CS and after cure (31 vs. 46.5 vs. 50%, p < 0.05). Patients with cured CS had more prevalence of dyslipidemia and central obesity than controls (p < 0.05). No statistical differences were observed between both groups regarding prevalence of diabetes and hypertension (fig. 1). Furthermore, patients with cured CS had higher total and trunk fat mass percentage and less lean body mass and bone mineral content than controls despite the same BMI (27.8 ± 6 vs. 26.5 ± 5 in controls) [10], indicating that despite disappearance of hypercortisolism, body composition abnormalities do not revert to normal. These findings suggest that the changes induced by sustained hypercortisolism are irreversible, and persist after endocrine cure. How long hypercortisolism should be present to induce these changes is currently unknown, but seems a strong argument to attain an earlier diagnosis of CS, before irreversible morbidity occurs.

Health-Related Quality of Life in Patients with Cushing's Syndrome

Using the generic SF-36 questionnaire, impairment of QoL was observed in patients with CS which was greater in active than in cured patients, and did not normalize, even after long-term control of hypercortisolism [11, 12]. With a recently described disease-generated questionnaire (CushingQoL) HRQoL was evaluated in 125 patients with CS in clinical practice conditions, recruited from Spain, France, Germany, The Netherlands and Italy, in an observational, international, cross-sectional study over a 2-month period [9]. Clinical and hormonal data were collected and correlated with results of a generic questionnaire (SF-36), a question on self-perceived general health status and the CushingQoL score. A significant correlation was observed between CushingQoL score and patients self-perceived general health status and dimensions of SF-36. Patients with current hypercortisolism scored worse (lower) than those without, and a linear regression analysis identified female gender and current hypercortisolism as significant predictors for worse QoL.

Conclusions

A high prevalence of cardiovascular risk factors is present in patients with active CS, and persists after long-term remission of hypercortisolism, mainly dyslipidemia and central obesity, as shown by our case-control study. Of special interest is the fact that despite a mean of cure of 11 years the prevalence of increased trunk fat and decreased bone mass persist, as in the active phase of CS. This persistent central obesity is a common complaint of patients exposed to hypercortisolism. Moreover, HRQoL is worse in patients with active disease, but is still impaired once the patients have been successfully treated for CS, compared to normal healthy populations. These findings may have practical implications, since it would be necessary to ascertain whether the same occurs in patients exposed to pharmacological doses of exogenous glucocorticoids.

Acknowledgements

Supported by grants from the Instituto de Salud Carlos III (FIS 05/0448). M.J. Barahona was supported by a fellowship from CIRIT (FI 03/1102).

References

1 Etxabe J, Vázquez JA: Morbidity and mortality in Cushing's disease: an epidemiological approach. Clin Endocrinol 1994;40:479–484.
2 Ambrosi B, Sartorio A, Pizzocaro A, Passini E, Bottasso B: Evaluation of haemostatic and fibrinolytic markers in patients with Cushing's syndrome and in patients with adrenal incidentaloma. Exp Clin Endocrinol Diabetes 2000;108:294–298.
3 Cavagnini F, Pecori-Giraldi F: Epidemiology and follow-up of Cushing's disease. Ann Med 2001;62:168–172.
4 Rees DA Hanna FW, Davies JS, Mills RG, Vafidis J, Scanlon: Long-term follow-up results of transsphenoidal surgery for Cushing's disease in a single centre using strict criteria for remission. Clin Endocrinol 2002;56:541–551.

5 Swearingen B, Biller BMK, Barker FG, Katznelson L, Grinspoon S, Klibanski A, Zervas NT: Long-term mortality after transsphenoidal surgery for Cushing's disease. Ann Intern Med 1999;245:821–824.
6 Pikkarainen L, Sane T, Reunanen A: The survival and well-being of patients treated for Cushing's syndrome. J Intern Med 1999;245:463–468.
7 Colao A, Pivonello R, Spiezia S, Faggiano A, Ferone D, Filippella M, Marzullo P, Cerbone G, Siciliani M, Lombardi G: Persistence of increased cardiovascular risk in patients with Cushing's disease after five years of successful cure. J Clin Endocrinol Metab 1999;84:2664–2672.
8 Van Aken MO, Pereira AM, Biermasz NR, van Thiel SW, Hoftijzer HC, Smit JW, Roelfsema F, Lamberts SW, Romijn JA: Quality of life in patients after long-term biochemical cure of Cushing's disease. J Clin Endocrinol Metab 2005;90:3279–3786.
9 Webb SM, Badia X, Barahona MJ, Colao A, Strasburger CJ, Tabarin A, van Aken MO, Pivonello R, Stalla G, Lamberts SWJ, Glusman JE: Evaluation of health-related quality of life in patients with Cushing's syndrome with a new questionnaire (CushingQoL). Eur J Endocrinol 2008;158:623–630.
10 Barahona MJ, Sucunza N, Resmini E, Fernández-Real JM, Ricard W, Moreno JM, Puig T, Farrerons J, Webb S: Persistent increase in body fat mass and inflammation despite long-term cured Cushings syndrome. J Clin Endocrinol Metab 2009;94:3365–3371.
11 Lindsay JR, Nansel T, Baid S, Gumowski J, Nieman LK: Long-term impaired quality of life in Cushing's syndrome despite initial improvement after surgical remission. J Clin Endocrinol Metab 2006;91:447–453.
12 Lindholm J, Juul S, Jorgensen JO, Astrup J, Bjerre P, Feldt-Rasmussen U, Hagen C, Jorgensen J, Kosteljanetz M, Kristensen L, Laurberg P, Schmidt K, Weeke J: Incidence and late prognosis of Cushing's syndrome: a population-based study. J Clin Endocrinol Metab 2001;86:117–123.

Susan M. Webb
Department of Endocrinology, Hospital Sant Pau
Pare Claret 167
ES–08025 Barcelona (Spain)
Tel. +34 93 5565661, Fax +34 93 5565602, E-Mail swebb@santpau.cat

Novel Medical Therapies for Pituitary Tumors

M. Theodoropoulou[a] · M. Labeur[a] · M. Paez Pereda[a,b] · M. Haedo[c] · M.J. Perone[c] · U. Renner[a] · E. Arzt[c] · G.K. Stalla[a]

[a]Max Planck Institute of Psychiatry, Department of Neuroendocrinology, and [b]Affectis Pharmaceuticals, Munich, Germany; [c]Laboratorio de Fisiología y Biología Molecular, Departamento de Fisiología y Biología Molecular y Celular, Facultad de Ciencias Exactas y Naturales, Universidad de Buenos Aires and IFIBYNE-CONICET, Buenos Aires, Argentina

Abstract

Despite considerable progress, there is still no medical treatment available for some kinds of pituitary tumors, in particular hormone inactive adenomas and corticotroph pituitary tumors. Surgical removal or at least debulking of the tumor is the only option to treat these kinds of tumors apart from rarely applied radiotherapy. Moreover, treatment resistance is present in a considerable proportion of patients bearing pituitary tumors, for which medical treatment regimens are already available (prolactinomas, somatotroph adenomas). Thus, novel or improved medical treatment strategies would be desirable. Here, we summarize preclinical and clinical findings about the hormone- and growth-suppressive action of various drugs, which will probably lead to novel future medical treatment concepts for pituitary tumors.

Copyright © 2010 S. Karger AG, Basel

Depending on the type of tumors, the primarily aim of medical treatment strategies for pituitary tumors is the normalization of excessive hormone secretion in endocrine-active microadenomas, the reduction of tumor size in nonfunctioning macroadenomas, and the achievement of both in hormone-secreting macroadenomas. Improvement of the medical therapy for pituitary tumors actually follows three directions namely the optimizing of already existing therapeutic strategies, the application of already used drugs in other types of pituitary tumors, and the development of completely new medical treatment concepts. For optimizing existing medical treatment strategies novel somatostatin analogues (e.g. pasireotide) with receptor subtype specificity have been developed as well as chimeric molecules, which target in parallel dopamine and somatostatin receptors. Based on findings about the expression of dopamine and somatostatin receptors in nonfunctioning and corticotroph adenomas, studies have been launched in which the efficacy of old and new somatostatin analogues and dopamine agonists is going to be studied in these kind of tumors. Ongoing studies of the pathomechanisms of pituitary tumors have led to the detection of novel potential drug targets such as components of intracellular signaling cascades. In the following an overview is given about the state of art of few of these extended or novel medical treatment concepts for pituitary adenomas.

Treatment of Non-Lactotroph Pituitary Tumors with Dopamine Agonists

Treatment of prolactinomas with dopamine agonists (bromocriptine, quinagolide, cabergoline) induces rapid normalization of prolactin levels and shrinkage of the tumors. Only in few patients, prolactinomas have to be removed by surgery due to resistance to dopamine treatment or because of compliance problems [1]. Since dopamine D2 receptors (D2R) have also been detected in the majority of nonfunctioning and corticotroph pituitary adenomas [2–6], the effects of dopamine agonists have been tested in these tumor types. After incomplete resection of nonfunctioning adenomas, subsequent dopamine agonist treatment mostly prevented regrowth or even induced further shrinkage of the remaining inactive tumor tissue [3, 4, 7]. Only in those nonfunctioning adenomas, which had been treated after re-manifestation of clinical symptoms, the efficacy of dopamine agonists was reduced [7, 8]. Thus, adjuvant therapy with dopamine agonist after partial surgical debulking of hormone inactive adenomas seems to be a promising future option in the medical treatment of this kind of tumor. Dopamine agonist treatment of patients with corticotroph adenomas for three months led to normalization of cortisol secretion in about 40% of the patients [5]. Therefore, for a subset of patients with Cushing´s disease medical therapy with dopamine agonists seems to be a future treatment option if other therapeutic approaches fail.

Novel Somatostatin Analogues in Pituitary Tumor Treatment

Pasireotide (SOM230)

Somatostatin has potent antisecretory and antiproliferative effects and therefore it has been a target for vigorous research for drug discovery [9]. Somatostatin analogues are the main medical treatment option for patients with various neuroendocrine tumors including gastroenteropancreatic and acromegaly-associated GH-secreting pituitary tumors. In acromegaly, the commonly used somatostatin analogs octreotide and lanreotide control GH and IGF-I in 50–60% of the cases, respectively, and induce tumor shrinkage in about 40%, indicating that approximately half of acromegaly patients remain uncured [10]. Somatostatin binds to a family of receptors (SSTR1–5), which belong to the seven-transmembrane-domain G-protein-coupled receptors [11]. Octreotide and lanreotide, primarily bind to SSTR2 and with less affinity to SSTR5.

Hence, there was intensive search for metabolically stable analogs mimicking the ability of the native somatostatin-14 to bind several SSTRs, which led to the synthesis of pasireotide (SOM230). Pasireotide binds SSTR1, 2, 3 and 5, displaying lower affinity for SSTR2 compared to octreotide and lanreotide, but higher to SSTR1, 3 and 5 [12]. These features render to pasireotide potent antisecretory and antitumor action in several neuroendocrine tumor models. In acromegalic patients, pasireotide displayed the same extend of GH suppression as octreotide, but it was also active in octreotide-resistant cases indicating that it could be a valuable pharmaceutical mean for the medical treatment of resistant acromegalic tumors [13].

The high affinity of pasireotide for SSTR5 indicated that it could control hormone secretion from corticotrophinomas, which were shown to have high levels of this SSTR but do not generally respond to octreotide or lanreotide. Indeed pasireotide was able to inhibit ACTH secretion in the majority of human corticotrophinomas in primary cell culture [14, 15]. The suppressive effect of pasireotide was not abolished by dexamethasone treatment as happens in the case of octreotide. Detailed investigation has revealed that SSTR2 is downregulated by dexamethasone treatment, while this is not the case with SSTR5 [16]. These in vitro data provide a mechanistic basis for the better suppressive effect of pasireotide on ACTH secretion. Pasireotide is currently in phase II clinical trial for the pharmaceutical treatment of Cushing's disease. The results are promising since it was found to decrease urinary free cortisol levels in most patients after 25 days of treatment [17].

Chimeric SSTR/D2R Compounds

Functional interaction of G-protein-coupled receptors from different families was repeatedly observed in several models. Interestingly, SSTR5 was found to hetero-oligomerize with D2R resulting in enhanced activity in terms of cAMP suppression [18]. This observation paved the way to the development of compounds that can be recognized by one or more SSTR and D2R. One of the first chimeric compounds developed, BIM-23A387, was shown to strongly suppress GH and PRL secretion from human mammosomatotrophinomas in primary cell culture compared to single SSTR2 or D2R analogue treatment [19]. A most recent analog with high affinity for SSTR2, 5 and D2R (BIM-23A760) had a strong antisecretory effect in GH-secreting tumors from patients partially resistant to standard octreotide treatment.

The higher potency of these chimeric compounds suggested that they could be beneficial for the treatment of pituitary tumor types that cannot respond to the classical SSTR2 or D2R analogs. Nonfunctioning pituitary adenomas (NFPA) express SSTR2 [20] and D2R [21], but they do not benefit from their antiproliferative action [21–23]. In an in vitro study involving four centers, BIM-23A760 exerted antiproliferative action in almost 60% of the NFPA in primary cell culture [24]. Because it was shown that postoperative dopamine agonist treatment in NFPA patients significantly associates with decreased prevalence of residual tumor growth [5], the higher potency of chimeric SSTR/D2R compounds such as BIM-23A760 could be useful as primary or adjunctive treatment option to surgery in NFPA patients.

Novel Treatment Options in Cushing's Disease: Retinoic Acid and Interferon-Gamma

Cushing's disease is a severe clinical condition caused by hypersecretion of corticosteroids due to excessive adrenocorticotrophin (ACTH) secretion from a pituitary adenoma [25]. New findings on the mechanisms, which are responsible for ACTH hypersecretion has enabled the identification of new targets, namely retinoic acid and interferon-γ (IFN-γ), which may be used for future treatment of ACTH-secreting pituitary adenomas.

Retinoic Acid

The biological effects of retinoic acid, broadly used for the prevention and treatment of different human cancers, are mediated by the nuclear receptors RAR (retinoic acid receptor) and RXR (retinoic X receptor) [26, 27]. In AtT-20 pituitary ACTH-secreting tumor cells, retinoic acid decreases ACTH secretion by inhibiting the transcriptional activity of the transcription factors AP1 and Nur on the POMC gene, which encodes ACTH [28]. Treatment of human corticotrophinomas in primary cell culture also resulted in the inhibition of ACTH production. The antiproliferative action and the inhibition of ACTH produced by retinoic acid in vitro were confirmed in vivo in experimental ACTH-secreting tumors in nude mice [28]. Recently, a randomized study using retinoic acid in dogs with Cushing's disease was performed [29]. Dogs were treated with 2 mg/kg body weight/day with isotretinoin *all-trans* retinoic acid for a period of 180 days. The control group received ketoconazole, an established treatment for Cushing's disease in humans and dogs. A significant reduction in plasma ACTH and α-MSH was observed along the time in the retinoic acid treated group. Moreover, the cortisol/creatinine urine ratio and the pituitary adenoma size were also significantly decreased in dogs under retinoic acid treatment. The survival time after initiation of treatment was significantly longer in the retinoid acid group compared to the control group. An improvement in different clinical signs, such as returning to estrus, food intake, skin appearance and hair loss, was observed after the treatment with retinoic acid. Thus, retinoic acid treatment resulted in the resolution of the clinical phenotype observed in Cushing's disease.

Retinoic acid treatment may represent a therapeutic option for the inhibition of ACTH and cortisol production, as well as tumor growth in patients with Cushing's disease.

Interferon-γ

IFN-γ, a cytokine exerting potent antitumorigenic effects in a variety of cancers, was recently shown to inhibit proliferation and ACTH production in tumoral pituitary cells [30]. In AtT-20 pituitary ACTH-secreting tumor cells, IFN-γ acting on its transmembrane receptor activates the receptor-associated Janus kinases (JAK 1 and 2), which allows the recruitment of the signal transducer and activator of transcription STAT1. An activated JAK-STAT1cascade is required for IFN-γ inhibitory action on POMC promoter activity. Moreover, factor-κ B (NF-κB) plays a crucial role in this inhibition. In agreement with the data obtained in AtT-20 cells, IFN-γ inhibits ACTH production in human pituitary adenoma cells from patients with Cushing's disease.

Thus, the development of therapeutic agents that target this novel IFN-γ/JAK-STAT1/NF-κB pathway might provide a valuable approach for treating Cushing's disease.

Conclusion and Perspectives

Progress in medical therapy of pituitary tumors is not only restricted to the improvement or extension of already existing treatment concepts with improved somatostatin analogs or dopamine agonists or the combination of both. Recent studies on drugs targeting intracellular signaling proteins or transcription factors involved in the regulation of hormone production and/or growth have shown promising results in vitro and in animals in vivo suggesting a future role of these compounds in the development of novel medical therapies for pituitary tumors. Other compounds already used in the treatment of different types of tumors such as the polyphenolic substance curcumin or the epidermal growth factor receptor antagonist gefitinib have also shown potent antitumorigenic activities in pituitary tumors in vitro [31, 32]. However, more work is needed to confirm these results in vivo before these drugs can be applied for the medical treatment of pituitary adenomas.

References

1. Mancini T, Casanueva FF, Giustina A: Hyperprolactinemia and prolactinomas. Endocrinol Metab Clin North Am 2008;37:67–99.
2. Renner U, Arzberger T, Pagotto U, Leimgruber S, Uhl E, Müller A, Lange M, Weindl A, Stalla GK: Heterogeneous dopamine D2 receptor subtype messenger ribonucleic acid expression in clinically nonfunctioning pituitary adenomas. J Clin Endocrinol Metab 1998;83:1368–1375.
3. Pivonelle R, Matrone C, Filippella M, Cavallo LM, Di Somma C, Cappabianca P, Colao A, Annunziato L, Lombardi G: Dopamine receptor expression and function in clinically nonfunctioning pituitary tumors: comparison with the effectiveness of cabergoline treatment. J Clin Endocrinol Metab 2004;89:1674–1683.
4. Petrossians P, Ronci N, Valdes Socin H, Kalife A, Stevenaert A, Bloch B, Tabarin A, Beckers A: ACTH silent adenoma shrinking under cabergoline. Eur J Endocrinol 2001;144:51–57.

5 Pivonello R, Ferone D, de Herder WW, Kros JM, De Caro ML, Arvigo M, Annunziato L, Lombardi G, Colao A, Hofland LJ, Lamberts SW: Dopamine receptor expression and function in corticotroph pituitary tumors. J Clin Endocrinol Metab 2004;89: 2452–2462.

6 De Bruin C, Hanson JM, Meij BP, Kooistra HS, Waaijers AM, Uitterlinden P, Lamberts SWJ, Hofland LJ: Expression and functional analysis of dopamine receptor subtype 2 and somatostatin receptor subtypes in canine cushing's disease. Endocrinology 2008;149:4357–4366.

7 Greenman Y, Tordjman K, Osher E, Veshchev I, Shenkerman G, Reider-Groswasser II, Shegev Y, Quaknine G, Stern N: Postoperative treatment of clinically nonfunctioning pituitary adenomas with dopamine agonists decreases tumour remnant growth. Clin Endocrinol 2005;63:39–44.

8 Lohmann T, Trantakis C, Biesold M, Prothman S, Guenzel S, Schober R, Paschke R: Minor tumour shrinkage in nonfunctioning pituitary adenomas by long-term treatment with the dopamine agonist cabergoline. Pituitary 2001;4:173–178.

9 Weckbecker G, Lewis I, Albert R, Schmid HA, Hoyer D, Bruns C: Opportunities in somatostatin research: biological, chemical and therapeutic aspects. Nat Rev Drug Discov 2003;2:999–1017.

10 Freda PU, Katznelson L, van der Lely AJ, Reyes CM, Zhao S, Rabinowitz D: Long-acting somatostatin analog therapy of acromegaly: a meta-analysis. J Clin Endocrinol Metab 2005;90:4465–4473.

11 Patel YC, Greenwood MT, Panetta R, Demchyshyn L, Niznik H, Srikant CB: The somatostatin receptor family. Life Sci 1995;57:1249–1265.

12 Lewis I, Bauer W, Albert R, Chandramouli N, Pless J, Weckbecker G, Bruns C: A novel somatostatin mimic with broad somatotropin release inhibitory factor receptor binding and superior therapeutic potential. J Med Chem 2003;46:2334–2344.

13 Schmid HA: Pasireotide (SOM230): development, mechanism of action and potential applications. Mol Cell Endocrinol 2008;286:69–74.

14 Hofland LJ, van der Hoek J, Feelders R, van Aken MO, van Koetsveld PM, Waaijers M, Sprij-Mooij D, Bruns C, Weckbecker G, de Herder WW, Beckers A, Lamberts SW: The multi-ligand somatostatin analogue SOM230 inhibits ACTH secretion by cultured human corticotroph adenomas via somatostatin receptor type 5. Eur J Endocrinol. 2005;152:645–654.

15 Batista DL, Zhang X, Gejman R, Ansell PJ, Zhou Y, Johnson SA, Swearingen B, Hedley-Whyte ET, Stratakis CA, Klibanski A: The effects of SOM230 on cell proliferation and adrenocorticotropin secretion in human corticotroph pituitary adenomas. J Clin Endocrinol Metab 2006;91:4482–4488.

16 Van der Hoek J, Waaijers M, van Koetsveld PM, Sprij-Mooij D, Feelders RA, Schmid HA, Schoeffter P, Hoyer D, Cervia D, Taylor JE, Culler MD, Lamberts SW, Hofland LJ: Distinct functional properties of native somatostatin receptor subtype 5 compared with subtype 2 in the regulation of ACTH release by corticotroph tumor cells. Am J Physiol Endocrinol Metab 2005;289:E278–E287.

17 Boscaro M, Ludlam WH, Atkinson B, Glusman JE, Petersenn S, Reincke M, Snyder P, Tabarin A, Biller BM, Findling J, Melmed S, Darby CH, Hu K, Wang Y, Freda PU, Grossman AB, Frohman LA, Bertherat J: Treatment of pituitary-dependent Cushing's disease with the multireceptor ligand somatostatin analog pasireotide (SOM230): a multicenter, phase II trial. J Clin Endocrinol Metab 2009;94:115–122.

18 Rocheville M, Lange DC, Kumar U, Patel SC, Patel RC, Patel YC: Receptors for dopamine and somatostatin: formation of hetero-oligomers with enhanced functional activity. Science. 2000;288:154–157.

19 Saveanu A, Lavaque E, Gunz G, Barlier A, Kim S, Taylor JE, Culler MD, Enjalbert A, Jaquet P: Demonstration of enhanced potency of a chimeric somatostatin-dopamine molecule, BIM-23A387, in suppressing growth hormone and prolactin secretion from human pituitary somatotroph adenoma cells. J Clin Endocrinol Metab 2002;87:5545–5552.

20 Greenman Y, Melmed S: Heterogeneous expression of two somatostatin receptor subtypes in pituitary tumors. J Clin Endocrinol Metab 1994;78:398–403.

21 Bevan JS, Webster J, Burke CW, Scanlon MF: Dopamine agonists and pituitary tumor shrinkage. Endocr Rev 1992;13:220–240.

22 De Bruin TW, Kwekkeboom DJ, Van't Verlaat JW, Reubi JC, Krenning EP, Lamberts SW, Croughs RJ: Clinically nonfunctioning pituitary adenoma and octreotide response to long term high dose treatment, and studies in vitro. J Clin Endocrinol Metab 1992;75:1310–1317.

23 Katznelson L, Oppenheim DS, Coughlin JF, Kliman B, Schoenfeld DA, Klibanski A: Chronic somatostatin analog administration in patients with alpha-subunit-secreting pituitary tumors. J Clin Endocrinol Metab 1992;75:1318–1325.

24 Florio T, Barbieri F, Spaziante R, Zona G, Hofland LJ, van Koetsveld PM, Feelders RA, Stalla GK, Theodoropoulou M, Culler MD, Dong J, Taylor JE, Moreau JP, Saveanu A, Gunz G, Dufour H, Jaquet P: Efficacy of a dopamine-somatostatin chimeric molecule, BIM-23A760, in the control of cell growth from primary cultures of human non-functioning pituitary adenomas: a multi-center study. Endocr Relat Cancer 2008;15:583–596.

25 Newell-Price J, Bertagna X, Grossman AB, Nieman LK: Cushing's syndrome. Lancet 2006;367:1605–1617.
26 Chambon P: A decade of molecular biology of retinoic acid receptors. FASEB J 1996;10:940–954.
27 Zang XK, Hoffmann B, Tran PB, Graupner G, Pfahl M: Retinoid X receptor is an auxiliary protein of thyroid hormone and retinoic acid receptors. Nature 1992;355:441–446.
28 Páez-Pereda M, Kovalovsky D, Hopfner U, Theodoropoulou M, Pagotto U, Uhl E, Losa M, Stalla J, Grübler Y, Missale C, Arzt E, Stalla GK: Retinoic acid prevents experimental Cushing syndrome. J Clin Invest 2001;108:1123–1131.
29 Castillo V, Giacomini D, Páez-Pereda M, Stalla J, Labeur M, Theodoropoulou M, Holsboer F, Grossman AB, Stalla GK, Arzt E: Retinoic acid as a novel medical therapy for Cushing's disease in dogs. Endocrinology 2006;147:4438–4444.
30 Labeur M, Refojo D, Wölfel B, Stalla J, Vargas V, Theodoropoulou M, Buchfelder M, Paez-Pereda M, Arzt E, Stalla GK: Interferon-gamma inhibits cellular proliferation and ACTH production in corticotroph tumor cells through a novel Janus kinases-signal transducer and activator of transcription 1/nuclear factor-kappa B inhibitory signaling pathway. J Endocrinol 2008;199:177–189.
31 Miller M, Chen S, Woodliff J, Kansra S: Curcumin (diferuloylmethane) nhibits cell proliferation, induces apoptosis, and decreases hormone levels and secretion in pituitary tumor cells. Endocrinology 2008;149:4158–4167.
32 Vlotides G, Siegel E, Donangelo I, Gutman S, Ren SG, Melmed S: Rat prolactinoma cell growth regulation by epidermal growth factor receptor ligands. Cancer Res 2008;68:6377–6386.

Dr. Ulrich Renner
Max Planck Institute of Psychiatry, Department of Neuroendocrinology
Kraepelinstrasse 10
DE–80804 Munich (Germany)
Tel. +49 89 30622 349, Fax +49 89 30622 605, E-Mail renner@mpipsykl.mpg.de

Medical Therapy of Cushing's Disease: Where Are We Now?

Krystallenia I. Alexandraki · Ashley B. Grossman

Department of Endocrinology, St Bartholomew's Hospital, London, UK

Abstract

The goals of ideal medical therapy for Cushing's disease should be to target the aetiology of the disorder, and thus surgery is the current 'gold standard' treatment. However, no effective drug that directly and effectively targets the adrenocorticotropin-secreting pituitary adenoma has been found to date, and treatments to control the hypercortisolaemic state by adrenal-based therapy are frequently used. Inhibitors of adrenal steroidogenesis, adrenolytic agents, compounds with neuromodulatory properties, and ligands of different nuclear hormone receptors involved in hypothalamo-pituitary regulation currently used have been reviewed. Ketoconazole and metyrapone can control hypercortisolaemic states, as well as mitotane in selective cases, depending on their side effects and frequent monitoring. The somatostatin analogue pasireotide and the dopamine agonist cabergoline, as well as their combination, show some therapeutic promise, while retinoic acid analogues should be further investigated in the pituitary-targeted medical therapy of Cushing's disease. Since a percentage of patients treated with surgery are not cured, or improve and subsequently relapse, there is an urgent need for effective medical therapies for this disorder. At present, only cabergoline and pasireotide are under active investigation, while adrenal steroidogenesis inhibitors are still the mainstay treatments for the control of the hypercortisolaemic state.

Copyright © 2010 S. Karger AG, Basel

Cushing's disease (CD) is the most frequent cause of endogenous hypercortisolaemia [1]. The short- and long-term consequences of hypercortisolism dictate the necessity for the normalisation of cortisol levels. Surgical removal of the adenoma still represents the best first-line treatment, which may be followed by radiotherapy in cases of surgical failure. Drugs are an alternative as monotherapy, but may also used in addition to radiotherapy or radiosurgery while awaiting their delayed effects, before surgery to reverse the metabolic consequences and poor healing of hypercortisolaemia, or in patients who cannot be submitted to surgical procedures because of co-morbidities or who are unwilling to receive other types of treatment [2].

Medical therapy is not considered the primary treatment in CD as in prolactinomas or acromegaly, since the ideal drug that directly and reliably targets the adrenocorticotropin (ACTH)-secreting pituitary adenoma has not been found. Variable compounds with neuromodulatory properties and nuclear hormone receptor ligands involved in hypothalamus-pituitary-adrenal regulation have been tested. On the other hand, compounds that target glucocorticoid synthesis or function have so far been used to control the deleterious effects of the hypercortisolaemic state [1, 2]. The present article will summarise the validity of the drugs in current or future potential use for CD.

Inhibitors of Cortisol Secretion

These compounds decrease cortisol levels by direct inhibition of steroidogenesis at one or more enzymatic steps, treating effectively hypercortisolism without affecting the underlying tumour or restoring the normal hypothalamus-pituitary-adrenal secretory dynamics. These drugs require escalation after an initial low dose to minimise their side effects; their dose may need to increase over time, since corticotroph tumours have a higher than normal set-point for cortisol negative feedback and ACTH secretion may increase in parallel with the fall of cortisol levels, although in practice this is less of a problem that might be expected. The inhibition can be partial or complete, and a 'block-and-replacement' scheme may be required. In either case, frequent monitoring is required to achieve an acceptable clinical and biochemical profile, and patients should be carefully instructed about adrenal insufficiency symptoms or symptomatic evidence of recurrence of hypercortisolism [3, 4].

Metyrapone blocks 11β-hydroxylase and has a rapid onset of action [5]. It has been used as monotherapy or in combination with other drugs, and/or following radiotherapy with good control of cortisol levels in both short- and long-term studies [6, 7]. It has been used successfully preoperatively, lowering mean cortisol levels [6] but it has only been considered useful as adjunctive treatment for CD by some authors but not by others [8, 9]. Metyrapone is given at an initial daily dose of 0.5–1 g, in three to four divided doses daily, and may be increased every few days to a maximal daily dose of 6 g [2]. However, since metyrapone results in increased ACTH secretion and hence increased androgenic precursors, acne and hirsutism constitute a common cause of treatment discontinuation in women [2, 4]. The appearance of hypertension, hypokalaemia and oedema due to mineralocorticoid precursors are not frequent side effects. Gastrointestinal effects and dizziness might be also associated with the inadvertent induction of adrenocortical insufficiency [4, 6].

Ketoconazole, an imidazole derivative, acts at a number of sites, inhibiting cytochrome P450 enzymes, with strongest effect on 17,20-lyase, and reported effects on side-chain cleavage (scc), 16α-hydroxylase, 17α-hydroxylase, 18-hydroxylase, and 11β-hydroxylase [10]. A possible extra-adrenal action at the pituitary level [11] has not been confirmed, but in general ACTH concentrations do not increase during long-term treatment with ketoconazole [2]. Furthermore, at high concentrations it has been shown to be an antagonist of the glucocorticoid receptor in cultured hepatoma cells [12] and to bind to glucocorticoid receptors in cytosolic preparations of human mononuclear cells [13]. Ketoconazole has been administered as monotherapy [14], adjuvant therapy in conjunction with radiation therapy [15], or in combination with other drugs such as octreotide [16], with satisfactory results. Treatment with ketoconazole is usually started at a daily dose of 400 mg and increased every three or more days to a maximum of 1.2 g daily in up to 3–4 divided doses [2]. Ketoconazole is not an option in achlorhydric patients and those treated with proton pump antagonists unless it is formulated locally in an acidic vehicle [3], since gastric acidity is necessary to metabolise it into the active compound. An idiosyncratic hepatic dyscrasia occurs in about 1 in 15,000 cases, which may occasionally be fatal or require liver transplantation [2, 4]; side effects such as gastrointestinal symptoms, skin rashes, irregular menses and reversible hepatic dysfunction do not necessarily require its discontinuation [4]. The reported decreased libido, impotence and gynaecomastia are the main cause of treatment discontinuation in men. Finally, teratogenicity contraindicates its use in pregnancy and its interaction with other medications because of the potent inhibitory effects on cytochrome P450 enzymes needs attentive evaluation [2, 4, 17]. Nevertheless, it is particularly useful in women improving hirsutism and in dyslipidaemia and decreasing cholesterol synthesis [2, 4]. A recent long-term clinical study supported its clinical value as safe and efficacious treatment in CD, particularly in patients for whom surgery is contraindicated or has to been delayed during the investigation of an occult adenoma [14].

Etomidate is another imidazole derivative which inhibits several enzymatic steps, the 11β-hydroxylase more potently, 17β-hydroxylase equally and 17,20-lyase less than ketoconazole [4, 18]; in higher concentrations it has an effect on cholesterol side-chain cleavage [18]. A bolus dose of 0.03 mg/kg iv followed by infusion of 0.1 mg/kg/h with a maximal dose of 0.3 mg/kg/h is currently recommended [2]. Etomidate has been considered useful since it can be administered intravenously resulting in rapid control of cortisol levels when oral therapy cannot be administered, as in cases of critically ill patients with CD [19].

Fluconazole is an antifungal azole derivative which is less toxic than ketoconazole, that can be administered in an intravenous form, and has been used in non ACTH-independent Cushing's syndrome (CS), but its value in ACTH-dependent CS has been doubted [20].

Mitotane (O,p'DDD) inhibits steroidogenesis at the level of side-chain cleavage, 11β- and 18-hydroxylase and 3β-hydroxysteroid dehydrogenase [3, 4]. If used at high doses its metabolite binds macromolecules in adrenocortical cell mitochondria, leading to their destruction and cellular necrosis, resulting over time in permanent glucocorticoid and mineralcorticoid replacement therapy [3]. It has been considered to be highly effective in the long-term suppression of hypercortisolism in patients with CD because of this adrenolytic action, and is used as an adjunct to radiation therapy or as monotherapy [21], with long remission after cessation of treatment [22]. Mitotane has a slow onset of action (weeks or months) and is often started at 250–500 mg nightly with slow escalation of the dose to 4–9 g/day even if higher doses have been used [2, 4]; it is important to regularly monitor drug levels because of its long half-life, and the significant gastrointestinal and neurological adverse effects or the fertility issues in women [3, 4, 17]; hypercholesterolaemia, hypouricaemia, gynaecomastia and prolonged bleeding time have been documented [4]. When glucocorticoid replacement therapy is required, hydrocortisone and prednisone are preferred because their metabolism is not induced by mitotane as is the case of dexamethasone, which instead has the advantage of a long half-life [17]. In addition, mitotane induces a rise in cortisol-binding globulin which renders the measurement of total cortisol unreliable, relying on plasma ACTH measurement or urinary free cortisol (UFC) to avoid an adrenocortical crisis [4]. Of note, lower doses reduce the incidence of side effects and combination with metyrapone in the early stages of therapy is an option [23]. Finally, the description of signs resembling Nelson's syndrome after mitotane administration should be carefully evaluated in CD [24].

Inhibition of Cortisol Function-Glucocorticoid Antagonist

Mifepristone (RU 486) is a steroid that binds competitively to the glucocorticoid, androgen and progestin receptors and inhibits the action of the endogenous ligands [4, 25]. Assessment of responsivity and dose monitoring is difficult and one should rely on clinical parameters, since the drug interferes with the glucocorticoid negative feedback at the hypothalamo-pituitary level inducing a rise in ACTH, and cortisol secretion, while the loss of feedback necessitates frequent imaging limiting its use as monotherapy in CD [25]. This fact limits its therapeutic value because of a drug-induced increase in ACTH and cortisol levels that may overcome the receptor blockade [25]. However, 1 patient with an ACTH-secreting pituitary macroadenoma, receiving high doses (up to 25 mg/kg daily) after radiation therapy, responded extremely well with remission of life-threatening clinical symptomatology [26]. Potential adverse effects after high-doses and long periods include endometrial hyperplasia, adrenal insufficiency and severe hypokalaemia attributed to excessive cortisol activation of the mineralocorticoid receptor, and may require treatment with spironolactone [25, 26].

Neuromodulatory Compounds and Ligands of Different Nuclear Hormone Receptors

Dopaminergic Agonists

The rationale for this treatment is the presence of the D2 receptor in the anterior and intermediate lobe of the pituitary gland showing variable and heterogeneous expression in 80–89% of all types of pituitary tumours [27–29]. However, no specific binding of a dopamine agonist was demonstrated in some corticotroph pituitary tumours [30], and the D2 receptor was not demonstrated by imaging studies in CD or Nelson's syndrome [31]. Dopaminergic modulation of ACTH secretion has been suggested to occur via regulation of hypothalamic CRH release in addition to direct inhibition of ACTH secretion by the corticotrophs [29, 32]. Bromocriptine has shown variable results in humans and very limited therapeutic potential [32, 33]. However, its beneficial effect on cyclical CD has been used as a diagnostic tool [34]. Cabergoline has also been investigated and reported to inhibit ACTH secretion in vitro in cases with D2-receptor-positive cells [2, 28, 29]. In small series with short- and long-term treatments with doses of 2.5–3.5 mg/week, there was a drop in UFC and serum cortisol levels, and tumour shrinkage [2, 28]. In a more recent study, it was demonstrated that cabergoline treatment at doses of 1–7 mg/week was effective in controlling cortisol secretion for at least one to two years in more than one-third of patients with CD after failure of surgical treatment [35]. The long-term consequences of dopamine agonists (specifically cabergoline) [36] needs further evaluation.

Somatostatin Analogues

The rationale for this treatment is the presence of somatostatin receptors (SSTRs) in human corticotroph adenomas, with the subtypes 1, 2 and 5 being the most frequently found; both SSTR2 and 5 seem to be implicated in the regulation of ACTH release, but SSTR5 is considered to be the predominant receptor [37–39]. Octreotide, a predominantly SSTR2-ligand having moderate affinity for SSTR5, has been proved to be virtually ineffective in patients with CD [40]. Pasireotide or SOM-230, a multiligand SST analogue, has high binding affinity to SSTR5 and 1, 2 and 3 subtypes with functional activity 30-, 11- and 158-fold higher on SSTR1, 3 and 5, respectively, and approximately seven-fold lower on SSTR2 compared with octreotide [41]. Glucocorticoids have a differential impact on the expression of the different SSTRs with SSTR2 being down-regulated and SSTR5 being resistant to corticosteroid modulation, explaining the loss of action of octreotide but not of pasireotide after glucocorticoid pre-treatment [42]. In line with this fact, in the hypercortisolaemic state of CD, pasireotide was shown to inhibit ACTH secretion more than octreotide [38] and to suppress cell proliferation in human corticotroph tumours [39]. The use of pasireotide was investigated in a phase II, open-label, single-arm, multicentre pilot study of 29 patients which received subcutaneously pasireotide 600μg twice daily for 15 days with reduction in UFC, serum cortisol and plasma ACTH levels [43]. Mild side effects such as nausea, abdominal pain and loose stools or diarrhoea were observed, as well as the well-known worsening of glycaemic control which occasionally required further treatment [43]. A randomised, double-blind, phase III study in the same population is on-going, while in a case report a women received pasireotide for 10 months 300 µg daily with considerable clinical and biochemical improvement [44]. Finally, the recent observation of a high co-expression of SSTR5 and D2 in the majority of human corticotroph adenomas studied [45] supports the use of the somatostatin-dopamine ligand dopastatin as a trial agent in CD [46].

The initial enthusiasm for peroxisome proliferator-activated receptor-gamma (PPAR-γ) agonists (rosiglitazone and pioglitazone) as promising agents specifically targeting pituitary tumours, based on in vitro and rodent models [47], was dampened by a consistent lack of effect on ACTH and cortisol levels in normal subjects or patients with CD after short- or long-term administration [29]. Additionally, a recent study using sensitive methodology revealed that PPAR-γ receptor is only poorly expressed in human pituitary tissue while no specific abnormality of PPAR-γ expression was detected in corticotroph tumours, describing poor immunocytochemical expression in both normal pituitary and pituitary adenomas with only weak cytoplasmic staining [48]. Futhermore, the antiproliferative effects of rosiglitazone were shown only at very high doses and these were not blocked by a specific PPAR-γ antagonist [48].

On the other hand, retinoic acid might be a promising alternative since it has been demonstrated to be effective in reducing ACTH and cortisol levels, and in the amelioration of clinical signs, prolongation of survival, and pituitary tumour shrinkage without obvious side effects in animal models [49, 50], and may have additional clinical benefits [29].

Conclusions

Currently, there is no effective medical therapy that directly and reliably targets the ACTH-secreting pituitary adenoma. However, there are encouraging developments such as pasireotide and cabergoline, while dopastatin may also show therapeutic promise in CD. It will also be of considerable clinical interest to design therapeutic trials with retinoic acid analogues. In the meantime therapy targeting the adrenal remains the mainstay of therapy. Ketoconazole and metyrapone (either alone or in combination) are the most frequently used drugs, and appear to be more effective and better tolerated than other adrenal inhibitors. However, the presence of hirsutism frequently precludes metyrapone therapy in women, while gynaecomastia or hypogonadism complicates the use of ketoconazole therapy in men. Etomidate remains an important option when intravenous administration is required for a rapid treatment of severely ill patients, while fluconazole needs to be further evaluated. Mitotane might be an alternative but the difficulty in monitoring and the adverse effects determine that its use be confined to the minority of patients who are intolerant or not responsive to the aforementioned treatments, and who are unsuitable for adrenalectomy. Finally, mifepristone has been used in limited cases, but further evaluation of its safety and follow-up methodologies is needed. Nonetheless, if patients remain intolerant or incompletely responsive to medical therapy, it is extremely important to remember that bilateral laparoscopic adrenalectomy will cause immediate remission of hypercortisolaemia when all else fails [2].

References

1 Makras P, Toloumis G, Papadogias D, Kaltsas GA, Besser M: The diagnosis and differential diagnosis of endogenous Cushing's syndrome. Hormones 2006;5:231–250.
2 Biller BM, Grossman AB, Stewart PM, Melmed S, Bertagna X, Bertherat J, Buchfelder M, Colao A, Hermus AR, Hofland LJ, Klibanski A, Lacroix A, Lindsay JR, Newell-Price J, Nieman LK, Petersenn S, Sonino N, Stalla GK, Swearingen B, Vance ML, Wass JA, Boscaro M: Treatment of adrenocorticotropin-dependent Cushing's syndrome: a consensus statement. J Clin Endocrinol Metab 2008;93:2454–2462.
3 Nieman LK: Medical therapy of Cushing's disease. Pituitary 2002;5:77–82.
4 Morris D, Grossman A: The medical management of Cushing's syndrome. Ann NY Acad Sci 2002;970:119–133.
5 Carballeira A, Fishman LM, Jacobi JD: Dual sites of inhibition by metyrapone of human adrenal steroidogenesis: correlation of in vivo and in vitro studies. J Clin Endocrinol Metab 1976;42:687–695.
6 Verhelst JA, Trainer PJ, Howlett TA, Perry L, Rees LH, Grossman AB, Wass JA, Besser GM: Short- and long-term responses to metyrapone in the medical management of 91 patients with Cushing's syndrome. Clin Endocrinol 1991;35:169–178.
7 Child DF, Burke CW, Burley DM, Rees LH, Fraser TR: Drug control of Cushing's syndrome: combined aminoglutethimide and metyrapone therapy. Acta Endocrinol (Copenh) 1976;82:330–341.
8 Orth DN: Metyrapone is useful only as adjunctive therapy in Cushing's disease (editorial). Ann Intern Med 1978;89:128–130.
9 Jeffcoate WJ, Rees LH, Tomlin S, Jones AE, Edwards CR, Besser GM: Metyrapone in long-term management of Cushing's disease. Br Med J 1977;2:215–217.
10 Engelhardt D, Weber MM, Miksch T, Abedinpour F, Jaspers C: The influence of ketoconazole on human adrenal steroidogenesis: incubation studies with tissue slices. Clin Endocrinol (Oxf) 1991;35:163–168.
11 Jimenez Reina L, Leal-Cerro A, Garcia J, Garcia-Luna PP, Astorga R, Bernal G: In vitro effects of ketoconazole on corticotrope cell morphology and ACTH secretion of two pituitary adenomas removed from patients with Nelson's syndrome. Acta Endocrinol (Copenh) 1989;121:185–190.
12 Loose DS, Stover EP, Feldman D: Ketoconazole binds to glucocorticoid receptors and exhibits glucocorticoid antagonist activity in cultured cells. J Clin Invest 1983;72:404–408.
13 Pardes E, De Yampey JE, Moses DF, De Nicola AF: Regulation of glucocorticoid receptors in human mononuclear cells: effects of glucocorticoid treatment, Cushing's disease and ketoconazole. J Steroid Biochem Mol Biol 1991;39:233–238.
14 Castinetti F, Morange I, Jaquet P, Conte-Devolx B, Brue T: Ketoconazole revisited: a preoperative or postoperative treatment in Cushing's disease. Eur J Endocrinol 2008;158:91–99.
15 Sonino N, Boscaro M, Paoletta A, Mantero F, Ziliotto D: Ketoconazole treatment in Cushing's syndrome: experience in 34 patients. Clin Endocrinol (Oxf) 1991;35:347–352.
16 Vignati F, Loli P: Additive effect of ketoconazole and octreotide in the treatment of severe adrenocorticotropin-dependent hypercortisolism. J Clin Endocrinol Metab 1996;81:2885–2890.
17 Gross BA, Mindea SA, Pick AJ, Chandler JP, Batjer HH: Medical management of Cushing disease. Neurosurg Focus 2007;23:E10.
18 Lamberts SW, Bons EG, Bruining HA, de Jong FH: Differential effects of the imidazole derivatives etomidate, ketoconazole and miconazole and of metyrapone on the secretion of cortisol and its precursors by human adrenocortical cells. J Pharmacol Exp Ther 1987;240:259–264.
19 Drake WM, Perry LA, Hinds CJ, Lowe DG, Reznek RH, Besser GM: Emergency and prolonged use of intravenous etomidate to control hypercortisolemia in a patient with Cushing's syndrome and peritonitis. J Clin Endocrinol Metab 1998;83:3542–3544.
20 Riedl M, Maier C, Zettinig G, Nowotny P, Schima W, Luger A: Long term control of hypercortisolism with fluconazole: case report and in vitro studies. Eur J Endocrinol 2006;154:519–524.
21 Schteingart DE, Tsao HS, Taylor CI, McKenzie A, Victoria R, Therrien BA: Sustained remission of Cushing's disease with mitotane and pituitary irradiation. Ann Intern Med 1980;92:613–619.
22 Luton JP, Mahoudeau JA, Bouchard P, Thieblot P, Hautecouverture M, Simon D, Laudat MH, Touitou Y, Bricaire H: Treatment of Cushing's disease by O,p'DDD: survey of 62 cases. N Engl J Med 1979;300:459–464.

23 Trainer PJ, Besser M: Cushing's syndrome: therapy directed at the adrenal glands. Endocrinol Metab Clin North Am 1994;23:571–584.
24 Lim MC, Tan YO, Chong PY, Cheah JS: Treatment of adrenal cortical carcinoma with mitotane: outcome and complications. Ann Acad Med Singapore 1990;19:540–544.
25 Johanssen S, Allolio B: Mifepristone (RU 486) in Cushing's syndrome. Eur J Endocrinol 2007;157:561–569.
26 Chu JW, Matthias DF, Belanoff J, et al: Successful long-term treatment of refractory Cushing's disease with high-dose mifepristone (RU 486). J Clin Endocrinol Metab 2001;86:3568–3573.
27 Stefaneanu L, Kovacs K, Horvath E, Buchfelder M, Fahlbusch R, Lancranjan L: Dopamine D2 receptor gene expression in human adenohypophysial adenomas. Endocrine 2001;14:329–336.
28 Pivonello R, Ferone D, de Herder WW, Kros JM, De Caro ML, Arvigo M, Annunziato L, Lombardi G, Colao A, Hofland LJ, Lamberts SW: Dopamine receptor expression and function in corticotroph pituitary tumors. J Clin Endocrinol Metab 2004;89:2452–2462.
29 Alexandraki KI, Grossman AB: Pituitary-targeted medical therapy of Cushing's disease. Expert Opin Investig Drugs 2008 17:669–677.
30 Cronin MJ, Cheung CY, Wilson CB, Jaffe RB, Weiner RI: [3H]Spiperone binding to human anterior pituitaries and pituitary adenomas secreting prolactin, growth hormone, and adrenocorticotropic hormone. J Clin Endocrinol Metab 1980;50:387–391.
31 Pirker W, Riedl M, Luger A, Czech T, Rössler K, Asenbaum S, Angelberger P, Kornhuber J, Deecke L, Podreka I, Brücke T: Dopamine D2 receptor imaging in pituitary adenomas using iodine-123-epidepride and SPECT. J Nucl Med 1996;37:1931–1937.
32 Bevan JS, Webster J, Burke CW, Scanlon MF: Dopamine agonists and pituitary tumor shrinkage. Endocr Rev 1992;13:220–240.
33 Adachi M, Takayanagi R, Yanase T, Sakai Y, Ikuyama S, Nakagaki H, Osamura Y, Sanno N, Nawata H: Cyclic Cushing's disease in long-term remission with a daily low dose of bromocriptine. Intern Med 1996;35:207–211.
34 Miller JW, Crapo L: The medical treatment of Cushing's syndrome. Endocr Rev 1993;14:443–458.
35 Pivonello R, De Martino MC, Cappabianca P, De Leo M, Faggiano A, Lombardi G, Hofland LJ, Lamberts SW, Colao A: The medical treatment of Cushing's disease: effectiveness of chronic treatment with the dopamine agonist cabergoline in patients unsuccessfully treated by surgery. J Clin Endocrinol Metab 2009;94:223–230.
36 Kars M, Pereira AM, Bax JJ, Romijn JA: Cabergoline and cardiac valve disease in prolactinoma patients: additional studies during long-term treatment are required. Eur J Endocrinol 2008;159:363–367.
37 Nielsen S, Mellemkjaer S, Rasmussen LM, Ledet T, Olsen N, Bojsen-Møller M, Astrup J, Weeke J, Jørgensen JO: Expression of somatostatin receptors on human pituitary adenomas in vivo and ex vivo. J Endocrinol Invest 2001;24:430–437.
38 Hofland LJ, van der Hoek J, Feelders R, van Aken MO, van Koetsveld PM, Waaijers M, Sprij-Mooij D, Bruns C, Weckbecker G, de Herder WW, Beckers A, Lamberts SW: The multi-ligand somatostatin analogue SOM230 inhibits ACTH secretion by cultured human corticotroph adenomas via somatostatin receptor type 5. Eur J Endocrinol 2005;152:645–654.
39 Batista DL, Zhang X, Gejman R, Ansell PJ, Zhou Y, Johnson SA, Swearingen B, Hedley-Whyte ET, Stratakis CA, Klibanski A: The effects of SOM230 on cell proliferation and adrenocorticotropin secretion in human corticotroph pituitary adenomas. J Clin Endocrinol Metab 2006;91:4482–4488.
40 Lamberts SW, Uitterlinden P, Klijn JM: The effect of the long-acting somatostatin analogue SMS 201-995 on ACTH secretion in Nelson's syndrome and Cushing's disease. Acta Endocrinol (Copenh) 1989;120:760–766.
41 Schmid HA, Schoeffter P: Functional activity of the multiligand analog SOM230 at human recombinant somatostatin receptor subtypes supports its usefulness in neuroendocrine tumors. Neuroendocrinology 2004;80(suppl 1):47–50.
42 van der Hoek J, Waaijers M, van Koetsveld PM, Sprij-Mooij D, Feelders RA, Schmid HA, Schoeffter P, Hoyer D, Cervia D, Taylor JE, Culler MD, Lamberts SW, Hofland LJ: Distinct functional properties of native somatostatin receptor subtype 5 compared with subtype 2 in the regulation of ACTH release by corticotroph tumor cells. Am J Physiol Endocrinol Metab 2005;289:E278–E287.
43 Boscaro M, Ludlam WH, Atkinson B, Glusman JE, Petersenn S, Reincke M, Snyder P, Tabarin A, Biller BM, Findling J, Melmed S, Darby CH, Hu K, Wang Y, Freda PU, Grossman AB, Frohman LA, Bertherat J: Treatment of pituitary dependent Cushing's disease with the multi-receptor ligand somatostatin analog pasireotide (SOM230): a multicenter, phase II trial. J Clin Endocrinol Metab 2009;94:115–122.
44 Cukier K, Tewari R, Kurth F, Schmid HA, Lai C, Torpy DJ: Significant response to pasireotide (SOM230) in the treatment of a patient with persistent, refractory Cushing's disease. Clin Endocrinol (Oxf) 2009;71:305–307.

45 de Bruin C, Pereira AM, Feelders RA, Romijn JA, Roelfsema F, Sprij-Mooij DM, van Aken MO, van der Lelij AJ, de Herder WW, Lamberts SW, Hofland LJ: Co-expression of dopamine and somatostatin receptor subtypes in corticotroph adenomas. J Clin Endocrinol Metab 2009;94:1118–1124.

46 Ren SG, Kim S, Taylor J, Dong J, Moreau JP, Culler MD, Melmed S: Suppression of rat and human growth hormone and prolactin secretion by a novel somatostatin/dopaminergic chimeric ligand. J Clin Endocrinol Metab 2003;88:5414–5421.

47 Heaney AP, Fernando M, Melmed S: PPAR-gamma receptor ligands: novel therapy for pituitary adenomas. J Clin Invest 2003;111:1381–1388.

48 Emery MN, Leontiou C, Bonner SE, Merulli C, Nanzer AM, Musat M, Galloway M, Powell M, Nikookam K, Korbonits M, Grossman AB: PPAR-gamma expression in pituitary tumours and the functional activity of the glitazones: evidence that any anti-proliferative effect of the glitazones is independent of the PPAR-gamma receptor. Clin Endocrinol (Oxf) 2006;65:389–395.

49 Páez-Pereda M, Kovalovsky D, Hopfner U, Theodoropoulou M, Pagotto U, Uhl E, Losa M, Stalla J, Grübler Y, Missale C, Arzt E, Stalla GK: Retinoic acid prevents experimental Cushing syndrome. J Clin Invest 2001;108:1123–1131.

50 Castillo V, Giacomini D, Páez-Pereda M, Stalla J, Labeur M, Theodoropoulou M, Holsboer F, Grossman AB, Stalla GK, Arzt E: Retinoic acid as a novel medical therapy for Cushing's disease in dogs. Endocrinology 2006;147:4438–4444.

Ashley Grossman, FMedSci
Professor of Neuroendocrinology
St. Bartholomew's Hospital
London EC1A 7BE (UK)
Tel. +44 207 6018343, Fax +44 207 6018306, E-Mail A.B.Grossman@qmul.ac.uk

Optimizing Acromegaly Treatment

Marcello D. Bronstein

Neuroendocrine Unit, Division of Endocrinology and Metabolism, Hospital das Clinicas, University of Sao Paulo Medical School, Sao Paulo, SP, Brazil

Abstract

Acromegaly is a rare chronic and disabling disease with many comorbidities leading to a mortality rate three times higher than in the normal population, mainly due to cardiovascular diseases. Based on epidemiological evidence, the mortality rate is normalized to normal population values when treatment brings serum GH levels to <2.5 µg/l and IGF-1 levels to normal age/gender values. Many efficacious therapeutic approaches are currently available to acromegaly control, namely pituitary surgery, radiotherapy and medical therapy, which encompasses dopamine agonists (DA), somatostatin analogs (SA; currently considered the gold standard treatment) and the GH-receptor antagonist pegvisomant (PEG-V). The efficacy and indications of each therapy depend on the tumor size and invasiveness, patient's condition and the local availability of each treatment modality. Treatment with SA analogs controls about two thirds of acromegalics, with recent data highly suggesting that this result can be enhanced by the combination of SA with PEG-V, or with the DA cabergoline. Moreover, surgical tumor debulking, even noncurative, might overcome the resistance of GH-secreting adenomas to SA. Therefore, these strategies may optimize acromegaly control, and potentially reduce treatment costs.

Copyright © 2010 S. Karger AG, Basel

Acromegaly, a chronic disabling disease with a prevalence of 50–60 cases per million population, is almost invariably caused by a GH-secreting pituitary adenoma, and rarely by eutopic or ectopic GHRH production. Elevated GH and IGF-1 are the hallmarks of this endocrine disturbance, resulting in soft tissue and skeletal growth and deformations, with cardiovascular, respiratory, neuromuscular and metabolic complications, as well as impairment of other pituitary functions. About 80% of the GH-secreting tumors are macroadenomas, many of them leading to mass-effect manifestations as visual impairment and headaches [1]. The mortality rate is three times higher than of the normal population, mainly due to cardiovascular diseases. Fortunately, diverse treatment approaches are available for acromegaly control which can bring the mortality rate back to those of the normal population [1]. There are controversies regarding the best biochemical marker of control. Normalization of mortality rate have been demonstrated by epidemiological studies when random GH

<2.5 μg/l are reached (probably <1.0 μg/l measured by modern sensitive immunoassays), though these levels do not necessarily indicate normalized GH secretion [2]. Although few data addressing the issue of mortality and age/gender related IGF-1 normalization is available, this peptide, measured by a well-validated assay with age- and gender-specific normative data, is also an excellent marker of treatment efficacy [2]. Goals in the treatment of acromegaly involve control of GH levels as tight as possible so bringing IGF-1 to normal levels, elimination of tumor mass effect and prevention of tumor growth, preservation or even improvement of other pituitary functions, and avoidance of recurrences. Pituitary surgery, radiotherapy, and medical treatment with somatostatin analogs (SA) and/or dopamine agonists (DA) are current available therapeutic modalities. Additionally, a recently developed drug with GH-receptor-blocking proprieties provides a unique mechanism of action, acting peripherally instead of in the tumor to block IGF-1 generation.

Trans-sphenoidal adenomectomy remains the preferred primary treatment for microadenomas, with cure rates of 70–90% in experienced hands. However, these results are not expected to be achieved by less-skilled surgeons, and microadenomas represent less that one fourth of GH-secreting tumors. Concerning macroadenomas, about 50% of intrasellar and less than 30% of expanding/invasive tumors are controlled by surgery alone [3]. Additionally, patients presenting with important cardiovascular, respiratory and/or metabolic comorbidities are at increased surgical risk. Therefore, an important subset of acromegalic patients are not expected to be cured by surgery or are unable to undergo such a procedure, needing, therefore, other therapeutic approaches.

Radiotherapy was largely used in the past, but has as drawbacks the delay of GH/IGF-1 normalization, hypopituitarism in about 50% of patients and, rarely, important complications as actinic brain necrosis, vascular damage, secondary neoplasia, and decrease in quality of life. The better efficacy and the decrease of side effects of stereotactic irradiation as gamma-knife and LINAC are still under investigation [4].

Medical therapy for acromegaly emerged in the last two decades as an important tool for disease control, especially for those patients not cured by surgery or in whom surgery is contraindicated or non expected to control the disease. Three classes of drugs are available for acromegaly: dopamine agonists, somatostatin analogs and GH receptor antagonist.

Dopamine Agonists

The rationale for their use is the paradoxical effect of dopamine in the tumoral somatotroph, blocking instead of stimulating GH secretion. However, they show limited efficacy. Bromocriptine is reported to control GH/IGF-1 secretion in about 10% of acromegalic patients, whereas cabergoline use shows better results, reaching 39% of the cases, mainly those with discrete serum GH elevation and/or prolactin (PRL) co-secretion [5]. The advantage of DA use is their oral administration and relatively low price. However, the limited efficacy and the recent issue of heart valve disease in the context of cabergoline use [6] must be kept in mind.

Somatostatin Analogs

Somatostatin (SST) action is mediated via five specific SST receptor subtypes (SSTR1–5), and exerts an inhibitory effect both in hormone secretion and tumor growth. SSTR2 and SSTR5 are the main subtypes involved in the regulation of GH secretion. About two thirds of acromegalic patients can be controlled by the commercially available SA octreotide (OCT) and lanreotide (LAN), which bind with high affinity mainly to SSTR2 and to a lesser extent to SSTR5 and SSTR3 [7]. A recent compilation by Murray and Melmed [8] comparing the long-lasting (once monthly) formulations of the two SA (OCT LAR and LAN Autogel) in terms of acromegaly control showed similar results: 65.3 vs. 59.5% based on GH <2.5 µg/l, 46.7 vs. 52.7% based on IGF-1 normalization, and 36 vs. 41.9% based on both criteria, respectively. Our data on OCT-LAR show similar results: 74% (GH <2.5 µg/l), 54% (normal IGF-1) and 41% (both) [9]. Additionally, SA are able to significantly reduce tumor size of GH-secreting adenomas: Melmed et al. [10] compiling data from 14 studies encompassing 424 acromegalic patients showed that 36.6% of patients receiving primary SA therapy for acromegaly experienced a significant (>20%) reduction in tumor size. Data from our group [9] and from Cozzi et al. [11] showed 76 and 82% of significant tumor shrinkage, respectively. The response to treatment is related to the expression of SSTR subtypes, so the best responders to SA are acromegalic patients whose tumors express mainly SSTR2, whereas adenomas of acromegalics partially or completely resistant to SA mainly express other SSTR subtypes or are presented with very low or no expression of all of them [7]. Therefore, the development of SSTR subtypes specific analogs may solve, at least partially, the current problem of resistance to OCT and LAN in about one third of acromegalics. As a matter of fact, many analogs, both 'universal' as SOM-203 (Pasireotide, Novartis) [12] and selective as BIM-23244 (Biomeasure/Ipsen) [13] have been tested in vitro and in clinical trials, but their clinical applicability is not established to date.

Growth Hormone Receptor Antagonist

A new treatment modality for acromegaly was achieved by the development of a GH receptor antagonist derived from the GH molecule by artificial mutagenesis, with increased half-life by pegylation. This drug, named pegvisomant (PEG-V), acts in the periphery, blocking the effects of GH on its target organs by binding to GH receptors and preventing their functional dimerization, thus blocking GH signal transduction with reduction of IGF-I production [14]. As PEG-V inhibits the action of GH but not its secretion, GH concentrations cannot be used to evaluate treatment efficacy. IGF-I is used as a surrogate marker, together with clinical parameters. PEG-V is administered subcutaneously at a daily dose of 10, 15 or 20 mg, depending on IGF-1 normalization. The drug is highly effective, as IGF-1levels normalize in more than 90% of patients [15], and is indicated for patients in whom surgery and SA fail. Based on the drug mechanism of action, tumor increase could be expected, although it was only observed in a few patients. As a matter of fact, the issue of tumor expansion under PEG-V treatment remains uncertain, as it could be drug related, but could also be derived from the SA withdrawal or even from the adenoma biology [16]. The major drawbacks of use are the need of daily injections, the occurrence of alterations of the liver enzymes in few cases [17], and lipohypertrophy in some patients, which regressed in all when the medication was discontinued [18].

Optimizing Acromegaly Treatment

Although virtually all acromegalic patients can be controlled by the above-mentioned treatments, emerging data show that combination of therapeutic procedures may increase the effectiveness and reduce the costs of therapy. Three of these approaches will be described.

Combined Somatostatin Analogs/Pegvisomant Therapy

Patients, mainly those not controlled by SA monotherapy, may exhibit glucose intolerance or even progression to overt diabetes mellitus [19], although a very recent meta-analysis suggests that modifications of glucose homeostasis induced by SSA may have an overall minor clinical impact in acromegaly [20]. On the other hand, the SA may control tumor growth or reduce its volume even without biochemical control [21]. Combination of SA and PEG-V treatment seems to be an attractive option because tumor suppression is combined with GH receptor blockade, and the GH receptor antagonist may improve glucose metabolism [22]. Feenstra et al. [23] assessed the efficacy of the combination of long-acting SA once monthly and PEG-V once weekly in 26 patients with active acromegaly for up to 42 weeks. The dose of PEG-V was increased until the IGF-I concentration became normal or until a weekly dose of 80 mg was reached. IGF-I reached normal concentrations in 18 of 19 (95%) patients who completed 42 weeks of treatment, with a median weekly dose of 60 mg PEG-V (range 40–80 mg). The same Rotterdam group revisited their data with a larger group of acromegalics (n = 32) and for an extended period (median 138 weeks, range 35–149 weeks). After dose finding, IGF-I remained within the normal range in all subjects with PEG-V administered once (n = 24) or twice (n = 8) weekly, on a total weekly dose of 60 mg (range 40–160 mg). No tumor increase was observed: in fact, in 13% of patients a regression in tumor size by more than 25% occurred [24]. Data concerning glucose metabolism with PEG-V addition to SA therapy show either improvement of no change. Alteration of the liver enzymes was also described in patients on SA/PEG-V therapy. Neggers et al. [25] assessed the long-term safety of the drug combination in 86 acromegalics for a mean of 29 months. Twenty-three (27%) patients showed dose-independent PEG-V-related transient liver enzyme elevations, which occurred only once during continuation of combination therapy, but discontinuation and rechallenge induced a second episode. Ten of these patients with liver enzyme elevations also suffered from diabetes mellitus. The occurrence of transient liver enzyme elevation with the combined therapy seems to be higher than that observed in PEG-V monotherapy: the German PEG-V observational study showed transaminase elevations in 16 of 263 patients (6.1%) that spontaneously resolved in 8 of them and promptly normalized in 5 patients who discontinued treatment [17]. The question if SA use or metabolic-associated disturbances may have influenced such a discrepancy remains to be solved.

The reduction in the injection frequency and in the mean weekly dose as compared to PEG-V monotherapy may be explained by at least two mechanisms: (1) as a consequence of endogenous GH reduction due to the SA therapy, less PEG-V is needed to normalize serum IGF-I levels, because it meets less wild-type GH to compete for GH-receptor binding [24]; (2) SA-induced increase of PEG-V concentration [26]. Notwithstanding, from a practical point of view, the SA/PEG-V association may improve the management of acromegalic patients not controlled by SA monotherapy with lower dose and less frequency of PEG-V injections as compared to the GH-antagonist drug monotherapy.

Combined Somatostatin Analogs/Cabergoline Therapy

Recent data point to a beneficial effect of addition of the DA cabergoline (CAB), 1.0–3.5 mg/week, to OCT or LAN in acromegalic patients uncontrolled during SA monotherapy, particularly those partially responsible to the SA drug and presenting with modestly elevated IGF-1 levels [27–30]. Regarding IGF-1 normalization, the effectiveness of combined treatment ranged from 42 to 56% of cases. Differently from CAB monotherapy, no significant correlation between IGF-I normalization and hyperprolactinemia was found. Moreover, in the patients operated on, no significant correlation between IGF-I normalization and positive ICH for PRL was observed (table 1).

The mechanism underlying the improvement of hormonal response to SA/CAB combined treatment remains unknown. One explanation could be the ligand-induced heterodimerization between SSTR and D2R. Hetero-oligomerizations between SSTRs or between SSTR and other G-protein-coupled receptors, such as D2R, have

Table 1. IGF-1 normalization after cabergoline addition to octreotide-LAR or lanreotide in acromegalic patients resistant to SA monotherapy

	IGF-1 normalization %	% URNL-IGF-1 pre-CAB	Correlation with serum PRL	Correlation with IHC+ for PRL
Cozzl et al. [27]	42	133 ± 16	no	no
Gatta et al. [29]	44	165 ± 21	no	no
Selvarajah et al. [28]	50	156 ± 25	no	–
Jallad and Bronstein [30]	56	197 ± 117	no	no

% URNL-IGF-1 = Percentage above the upper limit of IGF-1 normality; IHC = immunohistochemistry.

been reported some years ago by Patel's group [31]. An important functional consequence of the SSTR/DR2 interaction is a marked enhancement of signaling and altered pharmacological properties relative to the individually expressed receptor [31]. As a matter of fact, some recent in vitro studies evaluated the efficacy of chimeric SA-DA molecules (as BIM-23A387, Biomeasure/Ipsen) on suppression of GH and PRL secretion compared to selective agonists for single receptors [32, 33]. Although the maximal inhibition of GH release induced by the individual and combined sst2 and D2DR analogs and by BIM-23A387 was similar, the mean EC_{50} for GH suppression by the chimeric compound was 50 times lower than that of the sst2 and D2DR analogs, either individually or in combination, pointing to an enhanced potency of the chimeric molecule [33]. However, it is debatable whether or not the association of the two drugs, OCT-LAR/CAB, is capable of interacting in the same way as the chimeric molecules. Another possible mechanism for the greater potency of combined therapy may be due to the effect of SA on the pharmacokinetics of cabergoline, as it was already shown that OCT treatment prolongs small intestine and colonic transit time [34]. Potentially, this could increase the bioavailability and, consequently, improve cabergoline action.

Somatostatin Analogs and Partial Tumor Removal

It is well known that, in general, there is an inverse relationship between basal serum GH levels and the percentage of GH/IGF-1 control in acromegalics on SA treatment. Therefore, surgical tumor debulking, even noncurative, could improve the effectiveness of SA drugs in patients partially resistant to SA drugs. Previous studies addressing this issue did not show differences regarding GH and IGF-I normalization

Table 2. Literature data regarding acromegaly control on somatostatin analogs before and after partial surgical tumor removal

Author	n	Treatment	SA before surgery, %		SA after surgery, %	
			GH	IGF-I	GH	IGF-I
Petrossians et al. [37]	24	Oct, Oct-LAR, Lan SR, Lan Autogel	29	46	54	78
Colao et al. [38]	86	Oct-LAR, Lan SR, CAB	14	10	56	55
Jallad et al. [40]	11	Oct-LAR	36	0	64	82
Karavitaki et al. [39]	26	Lan SR	31	42	69	88

Oct = Octreotide; Oct-LAR = octreotide LAR; CAB = cabergoline; Lan SR = lanreotide SR; Lan = lanreotide.

between primary and postsurgical OCT-LAR treatment [9, 35, 36]. However, the caveat of these studies was the absence of an intrapatient comparison. Four recent studies have compared the response of primary and postsurgical SA therapy in the same acromegalic patient (table 2). Petrossians et al. [37] reported normalization of GH levels in 29 and 54% and of IGF-I levels in 46 and 78% of the cases before and after surgery, respectively. However, during the second SA course, many patients were treated with different drugs, drug formulations or doses and had different follow-up durations, when compared to primary SA treatment. The multicenter retrospective study by Colao et al. [38] reported GH normalization in 14 and 56% and normal IGF-I levels in 10 and 55% of the cases, before and after surgery, respectively. Caveats such as previous radiotherapy and different medical treatment regimens in some patients were balanced by another analysis, including nonirradiated patients only and the subset of patients under the same drug. Karavitaki et al. [39] reported 26 patients primarily treated with lanreotide SR for 4 months. After surgery, 20 patients exhibited biochemical remission; thus only 6 patients received postoperative lanreotide. Normalization of GH was observed in 31 and 69% and of IGF-I levels in 42 and 88% of the cases before and after surgery, respectively. In these three studies, all patients who received SA as the primary treatment were later submitted to surgery, regardless of IGF-I normalization during the primary treatment. We studied the outcome of secondary OCT-LAR treatment in 11 acromegalic patients with uncontrolled disease during OCT-LAR primary therapy and after surgery [40]. During primary SA therapy 36 and 0% of the patients achieved normal GH and IGF-1 levels, respectively. At the second (postsurgical) course of OCT-LAR, GH and IGF-I were controlled in

64 and 82% of the 11 patients, respectively. All patients were treated with the same SA dose and during a similar time span in both primary treatment and after unsuccessful surgical therapy.

Data from these four studies demonstrate that, in patients uncontrolled on primary SA therapy, surgical tumor debulking, even noncurative, may contribute to IGF-I normalization when the drug is reinstituted.

Conclusion

The association of the GH receptor antagonist PEG-V or the DA cabergoline, as well as partial surgical tumor removal, are therapeutic strategies with proven efficacy for patients inadequately controlled by SA monotherapy.

References

1 Melmed S: Acromegaly; in Melmed S (ed): The Pituitary. Oxford, Blackwell Science, 2002, pp 419–454.
2 Holdaway IM, Bolland MJ, Gamble GD: A meta-analysis of the effect of lowering serum levels of GH and IGF-I on mortality in acromegaly. Eur J Endocrinol 2008;159:89–95.
3 Nomikos P, Buchfelder M, Fahlbusch R: The outcome of surgery in 668 patients with acromegaly using current criteria of biochemical 'cure'. Eur J Endocrinol 2005;152:379–387.
4 Jallad RS, Musolino NR, Salgado LR, Bronstein MD: Treatment of acromegaly: Is there still a place for radiotherapy? Pituitary 2007;10:53–59.
5 Abs R, Verhelst J, Maiter D, Van Acker K, Nobels F, Coolens JL, Mahler C, Beckers A: Cabergoline in the treatment of acromegaly: a study in 64 patients. J Clin Endocrinol Metab 1998;83:374–378.
6 Steiger M, Jost W, Grandas F, Van Camp G: Risk of valvular heart disease associated with the use of dopamine agonists in Parkinson's disease: a systematic review. J Neural Transm 2009;116:179–191.
7 Bronstein MD: Acromegaly: molecular expression of somatostatin receptor subtypes and treatment outcome. Front Horm Res 2006;35:129–134.
8 Murray RD, Melmed S: A critical analysis of clinically available somatostatin analog formulations for therapy of acromegaly. J Clin Endocrinol Metab 2008;93:2957–2968.
9 Jallad RS, Musolino NR, Salgado LR, Bronstein MD: Treatment of acromegaly with octreotide-LAR: extensive experience in a Brazilian institution. Clin Endocrinol (Oxf) 2005;63:168–175.
10 Melmed S, Sternberg R, Cook D, Klibanski A, Chanson P, Bonert V, Vance ML, Rhew D, Kleinberg D, Barkan A: A critical analysis of pituitary tumor shrinkage during primary medical therapy in acromegaly. J Clin Endocrinol Metab 2005;90:4405–4410.
11 Cozzi R, Montini M, Attanasio R, Albizzi M, Lasio G, Lodrini S, Doneda P, Cortesi L, Pagani G: Primary treatment of acromegaly with octreotide LAR: a long-term (up to nine years) prospective study of its efficacy in the control of disease activity and tumor shrinkage. J Clin Endocrinol Metab 2006;91:1397–1403.
12 Schmid HA: Pasireotide (SOM230): development, mechanism of action and potential applications. Mol Cell Endocrinol 2008;286:69–74.
13 Saveanu A, Gunz G, Dufour H, Caron P, Fina F, Ouafik L, Culler MD, Moreau JP, Enjalbert A, Jaquet P: Bim-23244, a somatostatin receptor subtype 2- and 5-selective analog with enhanced efficacy in suppressing growth hormone (GH) from octreotide-resistant human GH-secreting adenomas. J Clin Endocrinol Metab 2001;86:140–145.
14 Higham CE, Trainer PJ: Growth hormone excess and the development of growth hormone receptor antagonists. Exp Physiol 2008;93:1157–1169.
15 van der Lely AJ, Hutson RK, Trainer PJ, Besser GM, Barkan AL, Katznelson L, Klibanski A, Herman-Bonert V, Melmed S, Vance ML, Freda PU, Stewart PM, Friend KE, Clemmons DR, Johannsson G, Stavrou S, Cook DM, Phillips LS, Strasburger CJ, Hackett S, Zib KA, Davis RJ, Scarlett JA, Thorner MO: Long-term treatment of acromegaly with pegvisomant, a growth hormone receptor antagonist. Lancet 2001;358:1754–1759.

16 Jimenez C, Burman P, Abs R, Clemmons DR, Drake WM, Hutson KR, Messig M, Thorner MO, Trainer PJ, Gagel RF: Follow-up of pituitary tumor volume in patients with acromegaly treated with pegvisomant in clinical trials. Eur J Endocrinol 2008;159:517–523.

17 Strasburger CJ, Buchfelder M, Droste M, Mann K, Stalla GK, Saller B, German Pegvisomant Investigators: Experience from the German pegvisomant observational study. Horm Res 2007;68(suppl 5):70–73.

18 Bonert VS, Kennedy L, Petersenn S, Barkan A, Carmichael J, Melmed S: Lipodystrophy in patients with acromegaly receiving pegvisomant. J Clin Endocrinol Metab 2008;93:3515–3518.

19 Baldelli R, Battista C, Leonetti F, Ghiggi MR, Ribaudo MC, Paoloni A, D'Amico E, Ferretti E, Baratta R, Liuzzi A, Trischitta V, Tamburrano G: Glucose homeostasis in acromegaly: effects of long-acting somatostatin analogues treatment. Clin Endocrinol (Oxf) 2003;59:492–499.

20 Mazziotti G, Floriani I, Bonadonna S, Torri V, Chanson P, Giustina A: Effects of somatostatin analogs on glucose homeostasis: a meta-analysis of acromegaly studies. J Clin Endocrinol Metab 2009;94:1500–1508.

21 Casarini AP, Pinto EM, Jallad RS, Giorgi RR, Giannella-Neto D, Bronstein MD: Dissociation between tumor shrinkage and hormonal response during somatostatin analog treatment in an acromegalic patient: preferential expression of somatostatin receptor subtype 3. J Endocrinol Invest 2006;29:826–830.

22 Jørgensen JO, Feldt-Rasmussen U, Frystyk J, Chen JW, Kristensen LØ, Hagen C, Ørskov H: Cotreatment of acromegaly with a somatostatin analog and a growth hormone receptor antagonist. J Clin Endocrinol Metab 2005;90:5627–5631.

23 Feenstra J, de Herder WW, ten Have SM, van den Beld AW, Feelders RA, Janssen JA, van der Lely AJ: Combined therapy with somatostatin analogues and weekly pegvisomant in active acromegaly. Lancet 2005;365:1644–1646.

24 Neggers SJ, van Aken MO, Janssen JA, Feelders RA, de Herder WW, van der Lely AJ: Long-term efficacy and safety of combined treatment of somatostatin analogs and pegvisomant in acromegaly. J Clin Endocrinol Metab 2007;92:4598–4601.

25 Neggers S, de Herder W, Janssen J, Feelders R, Van der Lely A: Combined treatment for acromegaly with long-acting somatostatin analogues and pegvisomant: long-term safety up to 4.5 years (median 2.2 years) of follow-up in 86 patients. Eur J Endocrinol 2009;160:529–533.

26 van der Lely AJ, Muller A, Janssen JA, Davis RJ, Zib KA, Scarlett JA, Lamberts SW: Control of tumor size and disease activity during co-treatment with octreotide and the growth hormone receptor antagonist pegvisomant in an acromegalic patient. J Clin Endocrinol Metab 2001;86:478–481.

27 Cozzi R, Attanasio R, Lodrini S, Lasio G: Cabergoline addition to depot somatostatin analogues in resistant acromegalic patients: efficacy and lack of predictive value of prolactin status. Clin Endocrinol (Oxf) 2004;61:209–215.

28 Selvarajah D, Webster J, Ross R, Newell-Price J: Effectiveness of adding dopamine agonist therapy to long-acting somatostatin analogues in the management of acromegaly. Eur J Endocrinol 2005;152:569–574.

29 Gatta B, Hau DH, Catargi B, Roger P, Tabarin A: Re-evaluation of the efficacy of the association of cabergoline to somatostatin analogues in acromegalic patients. Clin Endocrinol (Oxf) 2005;63:477–478.

30 Jallad RS, Bronstein MD: Optimizing medical therapy of acromegaly: beneficial effects of cabergoline in patients uncontrolled with octreotide-LAR. Neuroendocrinology 2009;90:82–92.

31 Rocheville M, Lange DC, Kumar U, Patel SC, Patel RC, Patel YC: Receptors for dopamine and somatostatin: formation of hetero-oligomers with enhanced functional activity. Science 2000;288:154–157.

32 Gruszka A, Ren SG, Dong J, Culler MD, Melmed S: Regulation of growth hormone and prolactin gene expression and secretion by chimeric somatostatin-dopamine molecules. Endocrinology 2007;148:6107–6114.

33 Saveanu A, Lavaque E, Gunz G, Barlier A, Kim S, Taylor JE, Culler MD, Enjalbert A, Jaquet P: Demonstration of enhanced potency of a chimeric somatostatin-dopamine molecule, BIM-23A387, in suppressing growth hormone and prolactin secretion from human pituitary somatotroph adenoma cells. J Clin Endocrinol Metab 2002;87:5545–5552.

34 Veysey MJ, Thomas LA, Mallet AI, Jenkins PJ, Besser GM, Wass JA, Murphy GM, Dowling RH: Prolonged large bowel transit increases serum deoxycholic acid: a risk factor for octreotide-induced gallbladder stones. Gut 1999;44:675–682.

35 Colao A, Ferone D, Marzullo P, Cappabianca P, Cirillo S, Boerlin V, Lancranjan I, Lombardi G: Long-term effects of depot long-acting somatostatin analog octreotide on hormone levels and tumor mass in acromegaly. J Clin Endocrinol Metab 2001;86:2779–2786.

36 Attanasio R, Baldelli R, Pivonello R, Grottoli S, Bocca L, Gasco V, Giusti M, Tamburrano G, Colao A, Cozzi R: Lanreotide 60 mg, a new long-acting formulation: effectiveness in the chronic treatment of acromegaly. J Clin Endocrinol Metab 2003;88: 5258–5265.

37 Petrossians P, Borges-Martins L, Espinoza C, Daly A, Betea D, Valdes-Socin H, Stevenaert A, Chanson P, Beckers A: Gross total resection or debulking of pituitary adenomas improves hormonal control of acromegaly by somatostatin analogs. Eur J Endocrinol 2005;152:61–66.

38 Colao A, Attanasio R, Pivonello R, Cappabianca P, Cavallo LM, Lasio G, Lodrini A, Lombardi G, Cozzi R: Partial surgical removal of growth hormone-secreting pituitary tumors enhances the response to somatostatin analogs in acromegaly. J Clin Endocrinol Metab 2006;91:85–92.

39 Karavitaki N, Turner HE, Adams CB, Cudlip S, Byrne JV, Fazal-Sanderson V, Rowlers S, Trainer PJ, Wass JA: Surgical debulking of pituitary macroadenomas causing acromegaly improves control by lanreotide. Clin Endocrinol (Oxf) 2008;68:970–975.

40 Jallad RS, Musolino NR, Kodaira S, Cescato VA, Bronstein MD: Does partial surgical tumour removal influence the response to octreotide-LAR in acromegalic patients previously resistant to the somatostatin analogue? Clin Endocrinol (Oxf) 2007;67: 310–315.

Marcello D. Bronstein
Neuroendocrine Unit, Division of Endocrinology and Metabolism, Hospital das Clinicas
University of Sao Paulo Medical School, Av.
Dr. Eneas de Carvalho Aguiar 155 oitavo andar, bloco 3, CEP 05403–000
Sao Paulo, SP 01406–100 (Brazil)
Tel. +55 11 3069 6293, Fax +55 11 3069 7694, E-Mail mdbronstein@uol.com.br

Vasoinhibins and the Pituitary Gland

Isabel Méndez[a,b] · Claudia Vega[a] · Miriam Zamorano[a] · Bibiana Moreno-Carranza[a] · Gonzalo Martínez de la Escalera[a] · Carmen Clapp[a]

[a]Instituto de Neurobiología, Universidad Nacional Autónoma de México, Campus UNAM-Juriquilla, Querétaro, Qro, and [b]Instituto Nacional de Ciencias Médicas y Nutrición Salvador Zubirán, México, D.F., México

Abstract

Vasoinhibins are a family of peptides that inhibit blood vessel growth, dilation, permeability, and survival. They are generated by the proteolytic cleavage of prolactin by cathepsin D, matrix metalloproteases, and bone morphogenic protein-1. Lactotropes within the anterior pituitary gland produce and release vasoinhibins. Hypothalamic neurons within the supraoptic and paraventricular nuclei also synthesize prolactin and process it to vasoinhibins that are released locally or at the neurohypophyseal endings. While both the anterior and posterior pituitaries may function as sources of circulating vasoinhibins, these peptides could act as local regulators of pituitary gland functions including neovascularization and neurohypophyseal hormone release.

Copyright © 2010 S. Karger AG, Basel

A 16-kDa fragment of prolactin (PRL), referred to as 16K-PRL, was discovered in the anterior pituitary gland three decades ago [1]. Based on its greater mammary mitogenic potency than PRL, 16K-PRL was initially thought to mediate the growth-promoting actions of the parent hormone [2]. While later work failed to confirm this hypothesis [3], it became clear that proteolytic cleavage yielded PRL derivatives with unique functional properties. A variety of PRL N-terminal fragments, collectively named vasoinhibins and produced by enzymatic proteolysis or by recombinant technology, was found to have inhibitory effects on blood vessels that were absent in full-length PRL [4]. In search of the endogenous vasoinhibins, the hypothalamo-posterior pituitary system was discovered as a new site for PRL synthesis and conversion to vasoinhibins [5]. The release of vasoinhibins by both the anterior and posterior pituitary gland implies that these peptides may act as hormones on peripheral blood vessels, but also as local factors in the regulation of pituitary gland vascularization and function. Here, we briefly summarize the current knowledge about the generation and putative functions of vasoinhibins in the anterior and posterior pituitary gland.

Vasoinhibins

PRL is proteolytically cleaved by cathepsin D [6, 7], matrix metalloproteases [8], and bone morphogenic protein-1 [9] into N-terminal fragments that range in molecular mass from 15 to 18 kDa. Also, recombinant vasoinhibins of 14 kDa [10] and 16 kDa [11] have been produced. Notably, a functional determinant of vasoinhibins may lie within a sequence of 14 hydrophobic amino acids located in the N-terminal region of the second helix of PRL [12]. This region displays antiangiogenic activity but is buried within the intact PRL, and it remains to be determined how this functional determinant would be exposed to exert its actions.

Vasoinhibins act on endothelial cells to inhibit the growth of new blood vessels from pre-existing vessels (angiogenesis) [see reviews, 4, 13]. They block growth factor-induced proliferation, migration, and tube formation of endothelial cells in culture and prevent neovascularization in the cornea, the retina, and the chick-embryo chorioallantoic membrane. In addition, vasoinhibins stimulate endothelial cell apoptosis causing the regression of blood vessels in vivo and in vitro; they also interfere with vasodilation of isolated vessels and with the vasodilation and vasopermeability of retinal blood vessels in situ [14]. The receptors mediating the effects of vasoinhibins have not been identified, but their signaling mechanisms include inhibition of the MAPK and the Ras-Tiam1-Pak1 signaling pathways, the stimulation of NFκB-mediated activation of initiator and effector caspases, and the blockage of endothelial nitric oxide synthase (eNOS) via the inhibition of intracellular Ca^{2+} mobilization and the activation of protein phosphatase 2A, causing eNOS dephosphorylation [4, 13–15].

The various effects of vasoinhibins imply their role as physiological inhibitors of blood vessel function. Recent studies have supported this hypothesis by showing the presence of vasoinhibins in the retina [16] and cartilage [8], where angiogenesis is highly restricted and absent, respectively. Because blocking the intraocular generation and action of vasoinhibins results in retinal angiogenesis and vasodilation [16], these peptides are emerging as natural inhibitors of ocular angiogenesis [see review, 17]. Notably, vasoinhibins were originally described in the anterior pituitary gland, a highly vascularized organ where abundant, highly permeable blood vessels enable hormone release into the blood stream, raising the question of the functional relevance of the peptides at this site.

Anterior Pituitary Vasoinhibins

Vasoinhibins are generated upon reduction of PRL cleaved at its large disulfide loop. Cleaved PRL has been detected in normal anterior pituitary glands of rats, mice and humans [18–20], and in human PRL-secreting pituitary adenomas [7]. The reduced cleaved forms, or vasoinhibins, were identified only recently as N-terminal fragments [7, 21]. This identification is important because a 16-kDa C-terminal fragment of

PRL, generated by cleavage with thrombin, is devoid of antiangiogenic activity [22]. Vasoinhibins are present in lysates and conditioned media of primary cultures of anterior pituitary cells, indicating their active synthesis and release [21]. The generation of vasoinhibins in the anterior pituitary gland is likely due to the cleavage of PRL by cathepsin D. Cathepsin D cleaves rat PRL in vitro by excision of the tripeptide Leu^{146}-Val^{147}-Trp^{148} between residues Tyr^{145} and Tyr^{149} [6], and a form of PRL cleaved after Tyr^{145} has been detected in rat adenohypophyseal cells [23]. Because cathepsin D is active at acidic pH values (<5.5), its PRL-cleaving action would be limited to acidic compartments within lactotropes, like the PRL secretory granules. An acid pH also occurs in the extracellular milieu of tumors [24]. Notably, cultured GH4C1 pituitary adenoma cells were shown to secrete cathepsin D able to generate vasoinhibins under acidic conditions [25]. Mimicking the tumor environment through exposure to hypoxia decreased the secretion of cathepsin D by GH4C1 cells, implying that a reduced generation of vasoinhibins may occur in the microenvironment of tumors, thus helping to create an angiogenic condition more favorable for tumor progression [25]. It should be noted that, in contrast to other types of tumors, pituitary adenomas are less vascularized than the surrounding normal pituitary tissue [26], suggesting that a dominant inhibition of angiogenesis in these tumors plays a role in their slow progression and failure to metastasize. It remains to be determined whether or not vasoinhibins are among the antiangiogenic factors influencing the phenotype and behavior of these tumors.

In any event, the production and release of vasoinhibins by endocrine cells supports their hormonal nature and is consistent with the presence of vasoinhibin-like proteins in the serum of rats and humans [see review, 4]. Of note, the systemic circulation represents an efficient delivery system of antiangiogenic factors to maintain the quiescent state of blood vessels that normally occurs in most adult tissues. The posterior pituitary gland may also contribute to the systemic release of vasoinhibins.

Posterior Pituitary Vasoinhibins

The characterization of antibodies raised against vasoinhibins led to the initial discovery of their presence in the hypothalamo-posterior pituitary system [5]. Posterior pituitary vasoinhibins were characterized as N-terminal fragments of PRL by their immunological and antiangiogenic properties [5, 27]. They were localized within magnocellular neurons of both the hypothalamic supraoptic nucleus (SON) and paraventricular nucleus (PVN) and their axonal projections to the posterior pituitary gland [5, 28]. They derived from locally produced PRL, as they continued to be detected in the SON and PVN of hypophysectomized rats [27] and PRL mRNA is expressed in both hypothalamic nuclei [29]. Consistent with vasoinhibins entering the neuronal secretory pathway, they were localized within secretory granules in the somas and terminals of the SON and PVN [28] and found to be released via a Ca^{2+}-dependent exocytotic mechanism from isolated posterior pituitary lobes and hypothalamo-posterior pituitary explants [27].

The functional relevance of vasoinhibins at this site is unclear. Besides being released into the systemic circulation, vasoinhibins may act locally within the hypothalamo-posterior pituitary system. Under physiological conditions, the vasculature in the central nervous system is basically in a dormant state except in specific pathologies that, through hypoxia, activate the expression of proangiogenic factors such as vascular endothelial growth factor (VEGF). Notably, magnocellular neurons in both the PVN and the SON express high levels of VEGF, which mediate the increase in angiogenesis in response to hyperosmotic stimuli [30]. It is proposed that osmotic stimulation promotes the release of vasopressin from the PVN and SON causing the constriction of local arterioles leading to hypoxia [30]. Although vasoinhibins would oppose local vascularization via their well-known antiangiogenic properties, they also promote vasoconstriction [15] and vasopressin release [31]. Stimulation of vasopressin release by vasoinhibins indicates that these peptides can influence vascular homeostasis not only through direct effects on endothelial cells, but also by activating neuroendocrine mechanisms. The functional implications of these interactions are under current investigation.

Conclusions and Perspectives

Vasoinhibins are emerging as natural inhibitors of blood vessel function. Their production and release by the anterior pituitary gland and by the hypothalamo-posterior pituitary gland system suggest that vasoinhibins serve as hormones and local regulators of the quiescent state of blood vessels. Studies aimed at altering the production and action of pituitary gland vasoinhibins could help understand the mechanisms underlying the phenotype of pituitary adenomas and the control of blood vessel homeostasis under health and disease.

Acknowledgements

We thank F. López-Barrera and G. Nava for their expert technical assistance, and D.D. Pless for critically editing the manuscript. Supported by the National Council of Science and Technology of Mexico (Grants 87015 and 44387) and UNAM (grant IN200509).

References

1 Mittra I: A novel 'cleaved prolactin' in the rat pituitary. I. Biosynthesis, characterization and regulatory control. Biochem Biophys Res Commun 1980; 95:1750.

2 Mittra I: A novel 'cleaved prolactin' in the rat pituitary. II. In vivo mammary mitogenic activity of its N-terminal 16K moiety. Biochem Biophys Res Commun 1980;95:1760.

3 Clapp C, Sears PS, Russell DH, Richards J, Levay-Young BK, Nicoll CS: Biological and immunological characterization of cleaved and 16K forms of rat prolactin. Endocrinology 1988;122:2892.

4 Clapp C, Aranda J, Gonzalez C, Jeziorski MC, Martinez de la Escalera G: Vasoinhibins: endogenous regulators of angiogenesis and vascular function. Trends Endocrinol Metab 2006;17:301.

5 Clapp C, Torner L, Gutierrez-Ospina G, Alcantara E, Lopez-Gomez FJ, Nagano M, Kelly PA, Mejia S, Morales MA, Martinez de la Escalera G: The prolactin gene is expressed in the hypothalamic-neurohypophyseal system and the protein is processed into a 14-kDa fragment with activity like 16-kDa prolactin. Proc Natl Acad Sci USA 1994;91:10384.

6 Baldocchi RA, Tan L, King DS, Nicoll CS: Mass spectrometric analysis of the fragments produced by cleavage and reduction of rat prolactin: evidence that the cleaving enzyme is cathepsin D. Endocrinology 1993;133:935.

7 Piwnica D, Touraine P, Struman I, Tabruyn S, Bolbach G, Clapp C, Martial JA, Kelly PA, Goffin V: Cathepsin D processes human prolactin into multiple 16K-like N-terminal fragments: study of their antiangiogenic properties and physiological relevance. Mol Endocrinol 2004;18:2522.

8 Macotela Y, Aguilar MB, Guzman-Morales J, Rivera JC, Zermeno C, Lopez-Barrera F, Nava G, Lavalle C, Martinez de la Escalera G, Clapp C: Matrix metalloproteases from chondrocytes generate an antiangiogenic 16 kDa prolactin. J Cell Sci 2006;119:1790.

9 Ge G, Fernandez CA, Moses MA, Greenspan DS: Bone morphogenetic protein 1 processes prolactin to a 17-kDa antiangiogenic factor. Proc Natl Acad Sci USA 2007;104:10010.

10 Clapp C, Martial JA, Guzman RC, Rentier-Delure F, Weiner RI: The 16-kilodalton N-terminal fragment of human prolactin is a potent inhibitor of angiogenesis. Endocrinology 1993;133:1292.

11 Struman I, Bentzien F, Lee H, Mainfroid V, D'Angelo G, Goffin V, Weiner RI, Martial JA: Opposing actions of intact and N-terminal fragments of the human prolactin/growth hormone family members on angiogenesis: an efficient mechanism for the regulation of angiogenesis. Proc Natl Acad Sci USA 1999;96:1246.

12 Nguyen NQ, Tabruyn SP, Lins L, Lion M, Cornet AM, Lair F, Rentier-Delrue F, Brasseur R, Martial JA, Struman I: Prolactin/growth hormone-derived antiangiogenic peptides highlight a potential role of tilted peptides in angiogenesis. Proc Natl Acad Sci USA 2006;103:14319.

13 Clapp C, Thebault S, Martinez de la Escalera G: Roles of prolactin and vasoinhibins in the regulation of vascular function in mammary gland. J Mammary Gland Biol Neoplasia 2008;13:55.

14 Garcia C, Aranda J, Arnold E, Thebault S, Macotela Y, Lopez-Casillas F, Mendoza V, Quiroz-Mercado H, Hernandez-Montiel HL, Lin SH, de la Escalera GM, Clapp C: Vasoinhibins prevent retinal vasopermeability associated with diabetic retinopathy in rats via protein phosphatase 2A-dependent eNOS inactivation. J Clin Invest 2008;118:2291.

15 Gonzalez C, Corbacho AM, Eiserich JP, Garcia C, Lopez-Barrera F, Morales-Tlalpan V, Barajas-Espinosa A, Diaz-Munoz M, Rubio R, Lin SH, Martinez de la Escalera G, Clapp C: 16K-prolactin inhibits activation of endothelial nitric oxide synthase, intracellular calcium mobilization, and endothelium-dependent vasorelaxation. Endocrinology 2004;145:5714.

16 Aranda J, Rivera JC, Jeziorski MC, Riesgo-Escovar J, Nava G, Lopez-Barrera F, Quiroz-Mercado H, Berger P, Martinez de la Escalera G, Clapp C: Prolactins are natural inhibitors of angiogenesis in the retina. Invest Ophthalmol Vis Sci 2005;46:2947.

17 Clapp C, Thebault S, Arnold E, Garcia C, Rivera JC, de la Escalera GM: Vasoinhibins: novel inhibitors of ocular angiogenesis. Am J Physiol Endocrinol Metab 2008;295:E772.

18 Shah GN, Hymer WC: Prolactin variants in the rat adenohypophysis. Mol Cell Endocrinol 1989;61:97.

19 Sinha YN, Gilligan TA, Lee DW, Hollingsworth D, Markoff E: Cleaved prolactin: evidence for its occurrence in human pituitary gland and plasma. J Clin Endocrinol Metab 1985;60:239.

20 Sinha YN, Gilligan TA: A cleaved form of prolactin in the mouse pituitary gland: identification and comparison of in vitro synthesis and release in strains with high and low incidences of mammary tumors. Endocrinology 1984;114:2046.

21 Cruz-Soto M, Cosío G, Jeziorski MC, Vargas-Barraso V, Aguilar MB, Carabez A, Berger P, Satfig P, Thebault S, Martínez de la Escalera G, Clapp C: Cathepsin D is a primary protease for the generation of adenohypophyseal vasoinhibins: cleavage occurs within the prolactin secretory granules. Endocrinology 2009;150:5446.

22 Khurana S, Liby K, Buckley AR, Ben-Jonathan N: Proteolysis of human prolactin: resistance to cathepsin D and formation of a nonangiostatic, C-terminal 16K fragment by thrombin. Endocrinology 1999;140:4127.

23 Andries M, Tilemans D, Denef C: Isolation of cleaved prolactin variants that stimulate DNA synthesis in specific cell types in rat pituitary cell aggregates in culture. Biochem J 1992;281:393.

24 Stubbs M, McSheehy PM, Griffiths JR, Bashford CL: Causes and consequences of tumour acidity and implications for treatment. Mol Med Today 2000;6:15.

25 Cosio G, Jeziorski MC, Lopez-Barrera F, De La Escalera GM, Clapp C: Hypoxia inhibits expression of prolactin and secretion of cathepsin-D by the GH4C1 pituitary adenoma cell line. Lab Invest 2003;83:1627.

26 Turner HE, Harris AL, Melmed S, Wass JA: Angiogenesis in endocrine tumors. Endocr Rev 2003;24:600.

27 Torner L, Mejia S, Lopez-Gomez FJ, Quintanar A, Martinez de la Escalera G, Clapp C: A 14-kilodalton prolactin-like fragment is secreted by the hypothalamo-neurohypophyseal system of the rat. Endocrinology 1995;136:5454.
28 Mejia S, Morales MA, Zetina ME, Martinez de la Escalera G, Clapp C: Immunoreactive prolactin forms colocalize with vasopressin in neurons of the hypothalamic paraventricular and supraoptic nuclei. Neuroendocrinology 1997;66:151.
29 Torner L, Nava G, Duenas Z, Corbacho A, Mejia S, Lopez F, Cajero M, Martinez de la Escalera G, Clapp C: Changes in the expression of neurohypophyseal prolactins during the estrous cycle and after estrogen treatment. J Endocrinol 1999;161:423.
30 Alonso G, Gallibert E, Lafont C, Guillon G: Intrahypothalamic angiogenesis induced by osmotic stimuli correlates with local hypoxia: a potential role of confined vasoconstriction induced by dendritic secretion of vasopressin. Endocrinology 2008; 149:4279.
31 Mejia S, Torner LM, Jeziorski MC, Gonzalez C, Morales MA, de la Escalera GM, Clapp C: Prolactin and 16K prolactin stimulate release of vasopressin by a direct effect on hypothalamo-neurohypophyseal system. Endocrine 2003;20:155.

Dr. Carmen Clapp
Instituto de Neurobiología, Universidad Nacional Autónoma de México (UNAM)
Campus UNAM-Juriquilla
Querétaro, Qro., 76230 (México)
Tel. +52 442 238 1028, Fax +52 442 234 0344, E-Mail clapp@unam.mx

Multiple Sources of Information for the Hypothalamus

A.J. van der Lelij

Erasmus University Medical Center, Rotterdam, The Netherlands

Abstract

The cross-talk between several parts of the gut and the brain involves the exchange of information that enables an individual to optimize metabolism and to adapt it to potentially huge variations in caloric intake during, e.g. fasting and eating. It also ensures that already parts of the gut downstream of the oropharynx are informed about what kind of food will arrive soon and what to do with it. These phenomena are largely mysteries to us, but some light is shed over these fields. This review will address some of the developments that illustrate how sophisticated these ancient mechanism of cross-talk are and what they could mean to us as highly developed 'modern' mammals.

Copyright © 2010 S. Karger AG, Basel

The anterior pituitary originates from the fetal gut and its close contact with the hypothalamus already illustrates that the brain and gut need a constant exchange of information in order to properly control food intake and metabolism. But also issues which are related to or totally dependent on it are under control of this collaboration. Examples of this are fertility, growth and sleep. This selective review is meant to stimulate and invite the reader to follow the developments that are presented closely as they will surely provide new insights into why we are what we are with regard to behavior, taste, smell and metabolism.

Sleep

Chronic sleep debt is becoming increasingly common and affects millions of people in more-developed countries. Sleep debt is currently believed to have no adverse effect on health. In a beautiful study by Spiegel et al. [1], the effect of sleep debt on metabolic and endocrine functions was investigated. They assessed carbohydrate metabolism, thyrotrophic function, activity of the hypothalamo-pituitary-adrenal axis, and sympathovagal balance in 11 young men after time in bed had been restricted to 4 h per night for 6 nights. They compared this sleep-debt condition with measurements taken at the end of a sleep-recovery period when participants were allowed 12 h in bed per night for 6 nights. They found that glucose tolerance was lower in the sleep-debt condition than in the fully rested condition. Apparently, sleep debt has a harmful impact on carbohydrate metabolism and endocrine function. The effects are similar to those seen in normal ageing and, therefore, sleep debt may increase the severity of age-related chronic disorders.

So apparently, obesity and sleep disorders are strongly connected. According to Vgontzas et al. [2], available data provide the basis for a meaningful phenotypic and pathophysiologic subtyping of obesity. One subtype is associated with emotional distress, poor sleep, fatigue, HPA axis 'hyperactivity,' and hypercytokinemia while the other is associated with nondistress, better sleep but more sleepiness, HPA axis 'normo- or hypoactivity,' and hypercytokinemia. This proposed subtyping may lead to novel, preventive and therapeutic strategies for obesity and its associated sleep disturbances.

Already in childhood, these phenomena might play an important role. Landhuis et al. [3] assessed the association between sleep time in childhood and adult BMI in a birth cohort. They found that shorter childhood sleep times were significantly associated with higher adult BMI values. This association remained after adjustment for adult sleep time and the potential confounding effects of early childhood BMI, childhood socioeconomic status, parental BMIs, child and adult television viewing, adult physical activity, and adult smoking. In logistic regression analyses, more sleep time during childhood was associated with lower odds of obesity at 32 years of age. This association was significant after adjustment for multiple potential confounding factors. Therefore, they concluded that sleep restriction in childhood increases the long-term risk for obesity. Ensuring that children get adequate sleep may be a useful strategy for stemming the current obesity epidemic.

Smell and Metabolism

Insect chemoreception is mediated by a large and diverse superfamily of seven-transmembrane domain receptors. These receptors were first identified in *Drosophila*, but have since been found in other insects, including mosquitoes and moths. Expression and functional analysis of these receptors have been used to identify receptor ligands and to map receptors to functional classes of neurons. Many receptors detect general odorants or tastants, whereas some detect pheromones. The noncanonical receptor Or83b, which is highly conserved across insect orders, dimerizes with odorant and pheromone receptors and is required for efficient localization of these proteins to dendrites of sensory neurons. These studies provide a foundation for understanding the molecular and cellular basis of olfactory and gustatory coding [4].

Smell is an ancient sensory system present in organisms from bacteria to humans. In the nematode *Caenorhabditis elegans*, gustatory and olfactory neurons regulate aging and longevity. Using the fruit fly, *Drosophila melanogaster*, Libert et al. [5] showed that exposure to nutrient-derived odorants can modulate life span and partially reverse the longevity-extending effects of dietary restriction. Furthermore, mutation of odorant receptor Or83b resulted in severe olfactory defects, altered adult metabolism, enhanced stress resistance, and extended life span. Their findings indicate that olfaction affects adult physiology and aging in *Drosophila*, possibly through the perceived availability of nutritional resources, and that olfactory regulation of life span is evolutionarily conserved.

Tatar et al. [6] had already shown that diet restriction (DR) extends adult survival in *Drosophila* only a bit longer than a dozen years ago. Limiting the amount of dietary yeast was sufficient to increase life span. In the short time since this initial observation, work with *Drosophila* has revealed several insights into the mechanisms of DR. It has also uncovered many unanticipated technical issues. In a beautiful review, they accumulated literature of DR in *Drosophila* from this methodological lens to distill four important results: yeast restriction alone is sufficient to increase survival; diet affects survival through two distinct physiological responses, starvation and longevity assurance; mortality has no memory of its past with respect to nutrition; the molecular operation of DR may involve processes of deacetylation via Sir-2 and Rpd-3. Finally, it remains unknown whether or not DR functions through insulin-related signaling.

Taste

In many sensory systems, stimulus sensitivity is dynamically modulated through mechanisms of peripheral adaptation, efferent input, or hormonal action. In this way, responses to sensory stimuli can be optimized in the context of both the environment and the physiological state of the animal. Although the gustatory system critically influences food preference, food intake and metabolic homeostasis, the mechanisms for modulating taste sensitivity are poorly understood. Shin et al. [7] reported that glucagon-like peptide-1 (GLP-1) signaling in taste buds modulates taste sensitivity in behaving mice. They found that GLP-1 is produced in two distinct subsets of mammalian taste cells, while the GLP-1 receptor is expressed on adjacent intragemmal afferent nerve fibers. GLP-1 receptor knockout mice show dramatically reduced taste responses to sweeteners in behavioral assays, indicating that GLP-1 signaling normally acts to maintain or enhance sweet taste sensitivity. A modest increase in citric acid taste sensitivity in these knockout mice suggests GLP-1 signaling may modulate sour taste, as well. Together, these findings suggest a novel paracrine mechanism for the regulation of taste function.

Tasting sweet food elicits insulin release prior to increasing plasma glucose levels, known as cephalic phase insulin release (CPIR). The characteristic of CPIR is that plasma insulin secretion occurs before the rise of the plasma glucose level. In a recent study, Tonosaki et al. [8] examined whether taste stimuli placed on the tongue could induce CPIR, using female Wistar rats and five basic taste stimuli: sucrose (sweet), sodium chloride (salty), HCl (sour), quinine (bitter) or monosodium glutamate (umami). Rats reliably exhibited CPIR to sucrose. Sodium chloride, HCl, quinine, or monosodium glutamate did not elicit CPIR. The non-nutritive sweetener saccharine elicited CPIR. However, starch, which is nutritive but non-sweet, did not elicit CPIR although rats showed a strong preference for starch which is a source of glucose. In addition, they studied whether CPIR was related to taste receptor cell activity. They carried out the experiment in rats with bilaterally cut chorda tympani nerves, one of the gustatory nerves. After sectioning, CPIR was not observed for sweet stimulation. From these results, they conclude that sweetness information conducted by this taste nerve provides essential information for eliciting CPIR.

Adipose tissue also seems to be under some sort of control by taste. It has recently been proposed that the peripheral taste organ is one of the targets for leptin. In lean mice, leptin selectively suppresses gustatory neural and behavioral responses to sweet compounds without affecting responses to other taste stimuli, whereas obese diabetic db/db mice with defects in leptin receptor lack this leptin suppression on sweet taste. Nakamura et al. [9] examined potential links between leptin and sweet taste in humans. A total of 91 nonobese subjects were used to determine recognition thresholds using a standard staircase methodology for various taste stimuli. They observed that the recognition thresholds for sweet compounds exhibited a diurnal variation from 08.00 to 22.00 h that parallels variation for leptin levels, with the lowest thresholds in the morning and the highest thresholds at night. This diurnal variation is sweet-taste selective, as it was not observed in thresholds for other taste stimuli (NaCl, citric acid, quinine, and monosodium glutamate). The diurnal variation for sweet thresholds in the normal feeding condition (three meals) was independent of meal timing and thereby blood glucose levels. Furthermore, when leptin levels were phase-shifted following imposition of one or two meals per day, the diurnal variation of thresholds for sweet taste shifted in parallel. This synchronization of diurnal variation in leptin levels and sweet taste recognition thresholds suggests a mechanistic connection between these two variables in humans.

Abstract taste perception plays a key role in determining individual food preferences and dietary habits. Individual differences in bitter, sweet, umami, sour, or salty taste perception may influence dietary habits, affecting nutritional status and nutrition-related chronic disease risk. In addition to these traditional taste modalities there is growing evidence that 'fat taste' may represent a sixth modality. Several taste receptors have been identified within taste cell membranes on the surface of the tongue, and they include the T2R family of bitter taste receptors, the T1R receptors associated with sweet and umami taste perception, the ion channels PKD1L3 and PKD2L1 linked to sour taste, and the integral membrane protein CD36, which is a putative 'fat taste' receptor. Additionally, epithelial sodium channels and a vanilloid receptor, TRPV1, may account for salty taste perception. Common polymorphisms in genes involved in taste perception may account for some of the interindividual differences in food preferences and dietary habits within and between populations. This variability could affect food choices and dietary habits, which may influence nutritional and health status and the risk of chronic disease. A review by Garcia-Bailo et al. [10] nicely summarized the present state of knowledge of the genetic variation in taste, and how such variation might influence food intake behaviors.

Recent studies have suggested that neuropeptides could play previously unrecognized functional roles in peripheral gustation. To date, two peptides, cholecystokinin and vasoactive intestinal peptide, have been localized to subsets of taste bud (TB) cells (TBC) and one, cholecystokinin, has been demonstrated to produce excitatory physiological actions. A study by Zhao et al. [11] extended the knowledge of neuropeptides in TBC in three significant ways. First, using techniques of immunocytochemistry and RT-PCR, evidence is presented for the expression of a third peptide, neuropeptide Y (NPY). Like other peptide expression patterns, NPY expression is circumscribed to a subset of cells within the taste bud. Second, using physiological studies, they demonstrate that NPY specifically enhances an inwardly rectifying potassium current via NPY-Y1 receptors. This action is antagonistic to the previously demonstrated inhibitory effect exerted by cholecystokinin on the same current, thus providing important clues to their signaling roles in the TB. Third, using the technique of double-labeled fluorescent immunocytochemistry, the relationship of three subsets of neuropeptide-expressing TB cells to one another was examined. Remarkably, NPY expressions, although fewer in number than either the cholecystokinin or vasoactive intestinal peptide subsets, overlapped 100% with either peptide. Collectively, these three observations transform previously suggestive roles of neuromodulation by peptides in TB cells to more concrete signaling pathways. The extensive colocalization of these peptides suggests they may be subject to similar presynaptic influences of release yet have antagonistic postsynaptic actions.

In conclusion, the interplay between the brain, its nerves plus neurotransmitters and the gut is intriguing and fascinating. Starting from smell and taste to gut hormones and insulin (sensitivity), all of these factors are more linked to each other than anyone ever could imagine. For sure, future research will even make us more speechless when we realize how little we know of everyday phenomena, ranging from sleep to eating.

References

1 Spiegel K, Leproult R, Van Cauter E: Impact of sleep debt on metabolic and endocrine function. Lancet 1999;354:1435–1439.
2 Vgontzas AN, Bixler EO, Chrousos GP, Pejovic S: Obesity and sleep disturbances: meaningful subtyping of obesity. Arch Physiol Biochem 2008;114: 224–236.
3 Landhuis CE, Poulton R, Welch D, Hancox RJ: Childhood sleep time and long-term risk for obesity: a 32-year prospective birth cohort study. Pediatrics 2008;122:955–960.
4 Dahanukar A, Hallem EA, Carlson JR: Insect chemoreception. Curr Opin Neurobiol 2005;15:423–430.

5 Libert S, Zwiener J, Chu X, Vanvoorhies W, Roman G, Pletcher SD: Regulation of Drosophila life span by olfaction and food-derived odors. Science 2007; 315:1133–1137.
6 Tatar M: Diet restriction in Drosophila melanogaster: design and analysis. Interdiscip Top Gerontol 2007;35:115–136.
7 Shin YK, Martin B, Golden E, Dotson CD, Maudsley S, Kim W, Jang HJ, Mattson MP, Drucker DJ, Egan JM, Munger SD: Modulation of taste sensitivity by GLP-1 signaling. J Neurochem 2008;106:455–463.
8 Tonosaki K, Hori Y, Shimizu Y: Relationships between insulin release and taste. Biomed Res 2007; 28:79–83.
9 Nakamura Y, Sanematsu K, Ohta R, Shirosaki S, Koyano K, Nonaka K, Shigemura N, Ninomiya Y: Diurnal variation of human sweet taste recognition thresholds is correlated with plasma leptin levels. Diabetes 2008;57:2661–2665.
10 Garcia-Bailo B, Toguri C, Eny KM, El-Sohemy A: Genetic Variation in Taste and Its Influence on Food Selection. OMICS 2009;13:69–80.
11 Zhao FL, Shen T, Kaya N, Lu SG, Cao Y, Herness S: Expression, physiological action, and coexpression patterns of neuropeptide Y in rat taste-bud cells. Proc Natl Acad Sci USA 2005;102:11100–11105.

Prof. Dr. A.J. van der Lelij
Erasmus MC, Dept. of Medicine, Room D-437
PO Box 2040
NL–3000 CA Rotterdam (The Netherlands)
Tel. +31 10 463 2862, Fax +31 10 463 3639, E-Mail a.vanderlelij@erasmusmc.nl

New Insights in Ghrelin Orexigenic Effect

Carlos Diéguez[a] · Kátia da Boit[b] · Marta G. Novelle[b] ·
Pablo B. Martínez de Morentin[b] · Rubén Nogueiras[b] · Miguel López[a]

[a]Department of Physiology, School of Medicine, University of Santiago de Compostela, Santiago de Compostela (A Coruña), and [b]CIBER Fisiopatología de la Obesidad y Nutrición (CIBERobn), Barcelona, Spain

Abstract

Ghrelin, a peptide hormone first discovered as the endogenous ligand of the growth hormone secretagogue receptor (GHS-R), is predominantly produced and released into the circulation by ghrelin cells (X/A-like) of the stomach fundus cells. Ghrelin has multiple actions in multiple tissues. In particular, it is the most potent known endogenous orexigenic peptide, and plays a significant role in glucose homeostasis: deletion of the genes encoding ghrelin and/or its receptor prevents high-fat diet from inducing obesity, increases insulin levels, enhances glucose-stimulated insulin secretion and improves peripheral insulin sensitivity. In addition to its already mentioned roles, ghrelin has other activities including stimulation of pituitary hormones secretion, regulation of gastric and pancreatic activity, modulation of fatty acid metabolism via specific control of AMP-activated protein kinase (AMPK), and cardiovascular and hemodynamic activities. In addition, modulation of cartilage and bone homeostasis, sleep and behavioral influences, and modulation of the immune system, as well as effects on cell proliferation, are other relevant actions of ghrelin. In this review, we summarize several aspects of ghrelin effects at hypothalamic level and their implications in the control of food intake and energy balance.

Copyright © 2010 S. Karger AG, Basel

Ghrelin Peptide in the Control of Food Intake and Body Weight Homeostasis

Ghrelin, a hormone produced in the stomach with orexigenic properties [1–6], has attracted enormous interest as a potential antiobesity therapeutic target [7, 8]. Chronic ghrelin administration promotes weight gain and adiposity in rodents [2], as well as increasing voluntary food intake in humans [9]. Assessment of circulating ghrelin levels has shown the existence of an inverse relationship with body weight [10], but in obese patients diagnosed with Prader-Willi syndrome increased levels of ghrelin are detected, which could explain their hyperphagia and increased body weight [11]. Furthermore, loss-of-function experiments using ghrelin knockout (KO) mice [12] or ghrelin receptor KO (GHS-R KO) mice [13] have shown that lack of ghrelin function protects against early-onset obesity. Altogether, these studies indicate that ghrelin might be an important signal for humans and rodents to prepare for meal initiation [14–16].

Ghrelin Acts on the Hypothalamus Regulating Food-Intake through the GHS-R1a

It is well established that the effect of ghrelin on food-intake is exerted through the growth hormone secretagogue receptor 1a (GHS-R1a), as in KO mice for this gene there is completely absence of response to exogenously administered ghrelin in terms of food-intake [17]. Furthermore, this receptor is expressed in hypothalamic nuclei which are well known by their involvement in the regulation of food intake and body weight homeostasis [5, 18–24]. The levels of the transcript coding for the ghrelin receptor are regulated in a nuclei-specific fashion by hormonal signals and nutritional status. Thus, in the hypothalamic arcuate nucleus (ARC) of the Zucker fatty rats and in fasted rats, GHS-R expression is significantly increased in comparison to controls [5]. A single leptin intracerebroventricular injection attenuated the fasting-induced increase in GHS-R but had no effect in fed rats 2 h after injection, whereas leptin infusion for 24 h or longer significantly decreased GHS-R expression in fed rats [5]. Ghrelin, but not GHRP-6, a synthetic growth hormone (GH) secretagogue, significantly increased GHS-R expression in the ARC of normal but not in dwarf rats [5, 19]. These results showed that the level of GHS-R expression in the ARC is reduced by leptin and increased by ghrelin and that the effect of ghrelin may be GH dependent [5].

Ghrelin Effects on Food Intake Are Mediated by the Orexigenic Neuropeptides NPY and AgRP

The GHS-R1a is expressed in neuropeptide Y/agouti-related peptide (NPY/AgRP) neurons in the arcuate nucleus of the hypothalamus (ARC) [5, 18–24], indicating that these set of neurons could be involved in ghrelin action. On the other hand, the lack of expression of the GHS-R in other hypothalamic regions, such as the lateral hypothalamus (LHA) [18], where a large number of orexigenic neurons are located [4, 25–40], suggested that this hypothalamic region was unlikely to be a target of the effects of ghrelin on food intake. In keeping with this, adult male rats, fed or fasted, treated centrally (intracerebroventicularly, ICV) with ghrelin showed increased AgRP and NPY expression in the ARC (fig. 1) [3, 4, 41]. In contrast, no change was demonstrated in the mRNA levels of the other neuropeptides (melanin concentrating hormone (MCH) and preprorexin) studied at any time evaluated [4]. To further characterize the role of ghrelin on the AgRP/NPY neurons, its effects

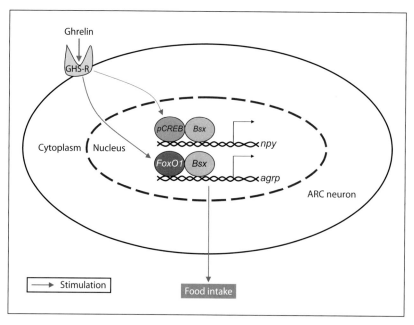

Fig. 1. Ghrelin effects on food-intake are mediated by the orexigenic neuropeptides NPY and AgRP, which are modulated by the transcription factor Bsx. Ghrelin, acting on GHS-R1a, modulates the expression of the homeobox domain transcription factor Bsx, stimulating its expression. In turn, Bsx promotes *npy* and *agrp* gene expression alongside phosphorylated cAMP response element binding protein (pCREB) and forkhead box O1 (FoxO1), respectively.

were studied in food-deprived animals. It is known that the nutritional status of the animals markedly influences the hypothalamic expression of neuropeptides involved in the regulation of food intake as well as their set-point and responsiveness to a vast array of stimuli [42–45]. As expected, we found that AgRP and NPY mRNA levels in the ARC were increased after 72 h of food deprivation. Taking into account that fasting increases ghrelin-circulating levels [2, 5, 6, 9], this suggests that the elevation of AgRP and NPY mRNA content could be mediated at least in part by ghrelin. This suggestion is reinforced by data showing that starvation-induced feeding was suppressed after central administration of anti-ghrelin antiserum [3]. It is noteworthy that even in this situation of increased circulating ghrelin levels, the administration of this peptide to food-deprived rats led to a further increase in AgRP and NPY mRNA contents in the ARC. The physiological relevance of both neuropeptides as mediators of ghrelin effects was firmly established by assessing the response to ghrelin in KO mice. These elegant experiments showed that while NPY KO or AgRP KO showed a normal response in terms of food intake to ghrelin, the double KO, NPY/AgRP failed to show any response, indicating the existence of redundancy among these two neuropeptides as mediators of ghrelin orexigenic action [46].

Transcription Factor Bsx Is Regulated by Ghrelin

It was recently reported that the hypothalamic homeobox domain transcription factor Bsx plays an essential role in the central nervous system controlling the spontaneous physical activity and the generation of hyperphagic responses [47, 48]. Moreover, it was found that Bsx is a master regulator for the hypothalamic expression of key orexigenic neuropeptides, namely NPY and AgRP (fig. 1). Therefore, we hypothesized that Bsx, which is expressed in the dorsomedial nucleus of the hypothalamus (DMH) and the ARC, could be regulated by afferent signals in response to peripheral energy balance such as ghrelin. This hypothesis was confirmed by the finding that ghrelin administration increased, whereas ghrelin receptor antagonist decreased, Bsx expression in the ARC (fig. 1) [48]. In contrast, leptin injection attenuated the fasting-induced increase in ARC Bsx levels but had no effect in fed rats [48]. Bsx expression in the DMH was unaffected by pharmacological modifications of leptin or ghrelin signaling [48]. Obese leptin-deficient (ob/ob) mice, but not obese melanocortin 4 receptor knockout mice, showed higher expression of Bsx, consistent with dependency from afferent leptin rather than increased adiposity per se [48]. Interestingly, exposure to a high-fat diet triggered Bsx expression, in agreement with the concept that decreased leptin signaling due to a high-fat diet increased Bsx expression [48]. Taken together, these data indicate that ARC Bsx expression is specifically regulated by afferent energy balance signals, including input from leptin and ghrelin, and that this transcription factor could be mediating the effects of ghrelin on NPY and AgRP (fig. 1) [47, 48].

Ghrelin Influences Hypothalamic Lipid Metabolism

Although the above-described findings allowed uncovering some of the mechanisms involved in the orexigenic effect of ghrelin, it was clearly apparent that we were still missing some of the key factors involved in the transduction pathway activated by binding of ghrelin to the GHS-R1a. In particular, it was unclear how the expression of these neuropeptides is regulated and by which specific signals and through what efferent systems they accurately match the bioenergetic needs of the organism. Current evidence suggests that nutrient-related metabolic pathways, such as fatty acid metabolism, may act as direct modulators of the hypothalamic control of feeding [6, 49–68], suggesting that perhaps ghrelin could exert its orexigenic effect through these metabolic routes. This hypothesis was fostered by recent data indicating that ghrelin modulates hypothalamic AMP-activated protein kinase (AMPK), a key upstream master regulator of lipid metabolism (fig. 2) [6, 53, 54, 62, 65, 66]. However, despite this evidence, the molecular mechanisms and anatomical details of this interaction were not been fully identified and, more importantly, there was no mechanistic data indicating that AMPK is required for ghrelin's orexigenic effects. We tested the possibility

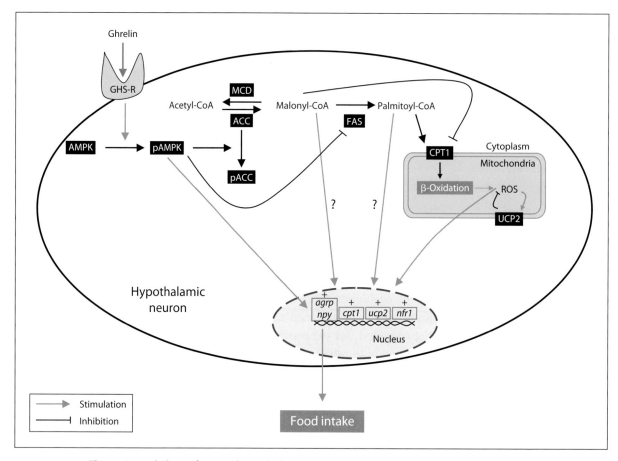

Fig. 2. Hypothalamic fatty acid metabolism integrates ghrelin with neuropeptide systems. Ghrelin, acting on the GHS-R1a, regulates AMPK, phosphorylating AMPK (pAMPK) and activating it, which in turn phosphorylates and inactivates acetyl-CoA carboxylase (ACC and pACC), decreasing the cytoplasmatic pool of malonyl-CoA and palmitoyl-CoA. This effect elicits changes in neuropeptide expression, such as AgRP and NPY by still undefined mechanisms (represented as ?), which ultimately regulates food intake. In addition, current evidence has demonstrated that ghrelin-induced carnitine palmitoyltransferase 1 (CPT1) activation promotes the generation of reactive oxygen species (ROS). Fatty acids and ROS increase uncoupling protein 2 (UCP2)-dependent uncoupling activity and UCP2 gene expression which subsequently decreases ROS in a feedback manner, allowing appropriate ghrelin-induced gene transcription. Further work is necessary to demonstrate whether this mechanism besides ghrelin, is extensive to other peripheral signals modulating food intake. NrF1 = Nuclear respiratory factor 1.

that fatty acid metabolism could be involved in ghrelin orexigenic action. By using a combination of pharmacological and genetic approaches, we first demonstrated that the physiological orexigenic response to ghrelin involves specific inhibition of fatty acid biosynthesis induced by AMPK resulting in decreased hypothalamic levels of malonyl-CoA and increased carnitine palmitoyltransferase-1 (CPT1) activity (fig. 2)

[6, 65]. In addition, we also demonstrate that fasting downregulates fatty acid synthase (FAS) in a region-specific manner and that this effect is mediated by an AMPK and ghrelin-dependent mechanisms (fig. 2) [6, 65]. Thus, decreasing AMPK activity in the ventromedial nucleus of the hypothalamus (VMH) is sufficient to inhibit ghrelin's effects on FAS expression and feeding [6, 65]. Taken together, these data provide novel evidence that the VMH is the hypothalamic nuclei where FAS is specifically downregulated in response to fasting [6, 57]. Interestingly, our results also indicate that the energy homeostasis-related signal sensed to regulate fatty acid metabolism in the hypothalamus and feeding behavior may not necessarily be a nutrient, but the stomach-secreted hormone ghrelin, acting through hypothalamic AMPK and that hypothalamic FAS expression is a downstream target of AMPK in the hypothalamus. The physiological significance of ghrelin effects on FAS-expressing neurons in the VMH is intriguing. The decrease in malonyl-CoA levels in response to ghrelin is transient so that by 6 h after ghrelin treatment, malonyl-CoA is restored to its normal levels [6]. Even though the mechanism by which malonyl-CoA levels are restored may be related to the termination of ghrelin stimulation of AMPK, our data also suggest that a direct effect of ghrelin decreasing FAS expression and activity in the VMH may also contribute to this process. Thus, we propose that ghrelin-induced and fasting-induced decreases in FAS levels in the VMH may be a physiological adaptive mechanism that helps to prevent malonyl-CoA from decreasing to deleteriously low levels in the hypothalamus [6, 65]. The observation that the decrease of FAS in response to fasting and ghrelin administration is limited to the VMH is particularly interesting. The VMH is well placed to integrate peripherally supplied signals that regulate food intake with the input from other neighboring nuclei and the brainstem, connected through specific neuronal projections [69]. Thus, our data provided evidence indicating that, besides its role of maintaining lipid biosynthesis, FAS in the VMH may also have an additional role as a sensor of the nutritional state [6, 65]. This possibility was further explored recently by others showing that the hypothalamic fatty acid oxidation pathway involving AMPK and CPT1 activated by ghrelin leads to a robust change in hypothalamic mitochondrial respiration and production of reactive oxygen species (ROS) in mice, which are dependent on uncoupling protein 2 (UCP2) (fig. 2) [66]. This activation of the mitochondrial mechanism is critical for ghrelin-induced mitochondrial proliferation and electric activation of NPY/AgRP neurons, for ghrelin-triggered synaptic plasticity of pro-opiomelanocortin-expressing (POMC) neurons and for ghrelin-induced food intake [66].

Concluding Remarks

In summary, we here show that the orexigenic action of ghrelin depends on hypothalamic GHS-R1a which is expressed in several neuronal populations well known for their involvement in the control of food intake. The work carried out by different groups over the last few years has uncovered the molecular mechanisms involved in ghrelin action. Noteworthy, the uncovering of a pathway involving AMPK-driven changes in hypothalamic lipid metabolism that subsequently influence neuropeptide gene expression possibly through the transcription factor Bsx offers several new drug targets that may be useful for the design of new molecules to be developed for the treatment of obesity.

Acknowledgements

This work was supported by the European Community (Health-F2–2008–223713, 'REPROBESITY'), and by Xunta de Galicia (ML: GRC2006/66), Fondo Investigationes Sanitarias (ML: PI061700), Ministerio de Educacion y Ciencia (ML: RyC-2007–00211 and RN: RYC-2008–02219), BFU 2008 (C.D. and M.L.). CIBER de Fisiopatología de la Obesidad y Nutrición is an initiative of ISCIII, Spain.

References

1 Kojima M, Hosoda H, Date Y, Nakazato M, Matsuo H, Kangawa K: Ghrelin is a growth-hormone-releasing acylated peptide from stomach. Nature 1999;402:656.
2 Tschop M, Smiley DL, Heiman ML: Ghrelin induces adiposity in rodents. Nature 2000;407:908.
3 Nakazato M, Murakami N, Date Y, Kojima M, Matsuo H, Kangawa K, Matsukura S: A role for ghrelin in the central regulation of feeding. Nature 2001;409:194.
4 Seoane LM, López M, Tovar S, Casanueva F, Señarís R, Diéguez C: Agouti-related peptide, neuropeptide Y, and somatostatin-producing neurons are targets for ghrelin actions in the rat hypothalamus. Endocrinology 2003;144:544.
5 Nogueiras R, Tovar S, Mitchell SE, Rayner DV, Archer ZA, Diéguez C, Williams LM: Regulation of growth hormone secretagogue receptor gene expression in the arcuate nuclei of the rat by leptin and ghrelin. Diabetes 2004;53:2552.
6 López M, Lage R, Saha AK, et al: Hypothalamic fatty acid metabolism mediates the orexigenic action of ghrelin. Cell Metab 2008;7:389.
7 Foster-Schubert KE, Cummings DE: Emerging therapeutic strategies for obesity. Endocr Rev 2006;27:779.
8 Zorrilla EP, Iwasaki S, Moss JA, Chang J, Otsuji J, Inoue K, Meijler MM, Janda KD: Vaccination against weight gain. Proc Natl Acad Sci USA 2006;103:13226.
9 Wren AM, Seal LJ, Cohen MA, Brynes AE, Frost GS, Murphy KG, Dhillo WS, Ghatei MA, Bloom SR: Ghrelin Enhances Appetite and Increases Food Intake in Humans. J Clin Endocrinol Metab 2001;86:5992.
10 Tschop M, Weyer C, Tataranni PA, Devanarayan V, Ravussin E, Heiman ML: Circulating ghrelin levels are decreased in human obesity. Diabetes 2001;50:707.
11 Cummings DE, Clement K, Purnell JQ, Vaisse C, Foster KE, Frayo RS, Schwartz MW, Basdevant A, Weigle DS: Elevated plasma ghrelin levels in Prader Willi syndrome. Nat Med 2002;8:643.
12 Wortley KE, Del Rincon JP, Murray JD, Garcia K, Iida K, Thorner MO, Sleeman MW: Absence of ghrelin protects against early-onset obesity. J Clin Invest 2005;115:3573.
13 Zigman JM, Nakano Y, Coppari R, Balthasar N, Marcus JN, Lee CE, Jones JE, Deysher AE, Waxman AR, White RD, Williams TD, Lachey JL, Seeley RJ, Lowell BB, Elmquist JK: Mice lacking ghrelin receptors resist the development of diet-induced obesity. J Clin Invest 2005;115:3564.
14 Tschop M, Wawarta R, Riepl RL, Friedrich S, Bidlingmaier M, Landgraf R, Folwaczny C: Postprandial decrease of circulating human ghrelin levels. J Endocrinol Invest 2001; 24:RC19–RC21.
15 Cummings DE, Purnell JQ, Frayo RS, Schmidova K, Wisse BE, Weigle DS: A preprandial rise in plasma ghrelin levels suggests a role in meal initiation in humans. Diabetes 2001;50:1714.

16 Drazen DL, Vahl TP, D'Alessio DA, Seeley RJ, Woods SC: Effects of a fixed meal pattern on ghrelin secretion: evidence for a learned response independent of nutrient status. Endocrinology 2006;147:23.

17 Sun Y, Wang P, Zheng H, Smith RG: Ghrelin stimulation of growth hormone release and appetite is mediated through the growth hormone secretagogue receptor. Proc Natl Acad Sci USA 2004;101:4679.

18 Guan XM, Yu H, Palyha OC, McKee KK, Feighner SD, Sirinathsinghji DJ, Smith RG, Van Der Ploeg LH, Howard AD: Distribution of mRNA encoding the growth hormone secretagogue receptor in brain and peripheral tissues. Brain Res Mol Brain Res 1997;48:23.

19 Bennett PA, Thomas GB, Howard AD, Feighner SD, Van Der Ploeg LH, Smith RG, Robinson IC: Hypothalamic growth hormone secretagogue-receptor (GHS-R) expression is regulated by growth hormone in the rat. Endocrinology 1997;138:4552.

20 Tannenbaum GS, Lapointe M, Beaudet A, Howard AD: Expression of growth hormone secretagogue-receptors by growth hormone-releasing hormone neurons in the mediobasal hypothalamus. Endocrinology 1998;139:4420.

21 Willesen MG, Kristensen P, Romer J: Co-localization of growth hormone secretagogue receptor and NPY mRNA in the arcuate nucleus of the rat. Neuroendocrinology 1999;70:306.

22 Mitchell V, Bouret S, Beauvillain JC, Schilling A, Perret M, Kordon C, Epelbaum J: Comparative distribution of mRNA encoding the growth hormone secretagogue-receptor (GHS-R) in *Microcebus murinus* (Primate, lemurian) and rat forebrain and pituitary. J Comp Neurol 2001;429:469.

23 Smith RG: Development of growth hormone secretagogues. Endocr Rev 2005;26:346.

24 Zigman JM, Jones JE, Lee CE, Saper CB, Elmquist JK: Expression of ghrelin receptor mRNA in the rat and the mouse brain. J Comp Neurol 2006;494:528.

25 Skofitsch G, Jacobowitz DM, Zamir N: Immunohistochemical localization of a melanin concentrating hormone-like peptide in the rat brain. Brain Res Bull 1985;15:635.

26 Risold PY, Fellmann D, Lenys D, Bugnon C: Coexistence of acetylcholinesterase-, human growth hormone-releasing factor(1–37)-, alpha-melanotropin- and melanin-concentrating hormone- like immunoreactivities in neurons of the rat hypothalamus: a light and electron microscope study. Neurosci Lett 1989;100:23.

27 Nahon JL, Presse F, Bittencourt JC, Sawchenko PE, Vale W: The rat melanin-concentrating hormone messenger ribonucleic acid encodes multiple putative neuropeptides coexpressed in the dorsolateral hypothalamus. Endocrinology 1989;125:2056.

28 Bittencourt JC, Presse F, Arias C, Peto C, Vaughan J, Nahon JL, Vale W, Sawchenko PE: The melanin-concentrating hormone system of the rat brain: an immuno- and hybridization histochemical characterization. J Comp Neurol 1992;319:218.

29 Qu D, Ludwig DS, Gammeltoft S, Piper M, Pelleymounter MA, Cullen MJ, Mathes WF, Przypek R, Kanarek R, Maratos-Flier E: A role for melanin-concentrating hormone in the central regulation of feeding behaviour. Nature 1996;380:243.

30 Sakurai T, Amemiya A, Ishii M, et al: Orexins and orexin receptors: a family of hypothalamic neuropeptides and G protein-coupled receptors that regulate feeding behavior. Cell 1998;92:573.

31 de Lecea L, Kilduff TS, Peyron C, et al: The hypocretins: hypothalamus-specific peptides with neuroexcitatory activity. Proc Natl Acad Sci USA 1998;95:322.

32 Broberger C, de Lecea L, Sutcliffe JG, Hokfelt T: Hypocretin/orexin- and melanin-concentrating hormone-expressing cells form distinct populations in the rodent lateral hypothalamus: relationship to the neuropeptide Y and agouti gene-related protein systems. J Comp Neurol 1998;402:460.

33 Peyron C, Tighe DK, van den Pol AN, de Lecea L, Heller HC, Sutcliffe JG, Kilduff TS: Neurons containing hypocretin (orexin) project to multiple neuronal systems. J Neurosci 1998;18:9996.

34 Elias CF, Saper CB, Maratos-Flier E, Tritos NA, Lee C, Kelly J, Tatro JB, Hoffman GE, Ollmann MM, Barsh GS, Sakurai T, Yanagisawa M, Elmquist JK: Chemically defined projections linking the mediobasal hypothalamus and the lateral hypothalamic area. J Comp Neurol 1998;402:442.

35 Nambu T, Sakurai T, Mizukami K, Hosoya Y, Yanagisawa M, Goto K: Distribution of orexin neurons in the adult rat brain. Brain Res 1999;827:243.

36 Date Y, Ueta Y, Yamashita H, Yamaguchi H, Matsukura S, Kangawa K, Sakurai T, Yanagisawa M, Nakazato M: Orexins, orexigenic hypothalamic peptides, interact with autonomic, neuroendocrine and neuroregulatory systems. Proc Natl Acad Sci USA 1999;96:748.

37 Horvath TL, Diano S, van den Pol AN: Synaptic interaction between hypocretin (orexin) and neuropeptide Y cells in the rodent and primate hypothalamus: a novel circuit implicated in metabolic and endocrine regulations. J Neurosci 1999;19:1072.

38 López M, Seoane L, García MC, Lago F, Casanueva FF, Señarís R, Diéguez C: Leptin regulation of prepro-orexin and orexin receptor mRNA levels in the hypothalamus. Biochem Biophys Res Commun 2000;269:41.

39 Chou TC, Lee CE, Lu J, Elmquist JK, Hara J, Willie JT, Beuckmann CT, Chemelli RM, Sakurai T, Yanagisawa M, Saper CB, Scammell TE: Orexin (hypocretin) neurons contain dynorphin. J Neurosci 2001;21:RC168.

40 López M, Seoane LM, García MC, Diéguez C, Señarís R: Neuropeptide Y, but not agouti-related peptide or melanin-concentrating hormone, is a target Peptide for orexin-a feeding actions in the rat hypothalamus. Neuroendocrinology 2002;75:34.

41 Kamegai J, Tamura H, Shimizu T, Ishii S, Sugihara H, Wakabayashi I: Chronic central infusion of ghrelin increases hypothalamic neuropeptide Y and Agouti-related protein mRNA levels and body weight in rats. Diabetes 2001;50:2438.

42 Abizaid A, Gao Q, Horvath TL: Thoughts for food: brain mechanisms and peripheral energy balance. Neuron 2006;51:691.

43 Morton GJ, Cummings DE, Baskin DG, Barsh GS, Schwartz MW: Central nervous system control of food intake and body weight. Nature 2006;443:289.

44 López M, Tovar S, Vázquez MJ, Williams LM, Diéguez C: Peripheral tissue-brain interactions in the regulation of food intake. Proc Nutr Soc 2007; 66:131.

45 Coll AP, Farooqi IS, O'Rahilly S: The hormonal control of food intake. Cell 2007;129:251.

46 Chen HY, Trumbauer ME, Chen AS, Weingarth DT, Adams JR, Frazier EG, Shen Z, Marsh DJ, Feighner SD, Guan XM, Ye Z, Nargund RP, Smith RG, Van Der Ploeg LH, Howard AD, MacNeil DJ, Qian S: Orexigenic action of peripheral ghrelin is mediated by neuropeptide Y (NPY) and agouti-related protein (AgRP). Endocrinology 2004;145:2607.

47 Sakkou M, Wiedmer P, Anlag K, Hamm A, Seuntjens E, Ettwiller L, Tschop MH, Treier M: A role for brain-specific homeobox factor bsx in the control of hyperphagia and locomotory behavior. Cell Metab 2007;5:450.

48 Nogueiras R, López M, Lage R, Pérez-Tilve D, Pfluger P, Mendieta-Zeron H, Sakkou M, Wiedmer P, Benoit S, Datta R, Dong JZ, Culler M, Sleeman M, Vidal-Puig A, Horvath T, Treier M, Diéguez C, Tschop MH: Bsx, a novel hypothalamic factor linking feeding with locomotor activity, is regulated by energy availability. Endocrinology 2008;49:3009.

49 Loftus TM, Jaworsky DE, Frehywot GL, Townsend CA, Ronnett GV, Lane MD, Kuhajda FP: Reduced food intake and body weight in mice treated with fatty acid synthase inhibitors. Science 2000;288:2379.

50 Hu Z, Cha SH, Chohnan S, Lane MD: Hypothalamic malonyl-CoA as a mediator of feeding behavior. Proc Natl Acad Sci USA 2003;100:12624.

51 Obici S, Feng Z, Arduini A, Conti R, Rossetti L: Inhibition of hypothalamic carnitine palmitoyl-transferase-1 decreases food intake and glucose production. Nat Med 2003;9:756.

52 Minokoshi Y, Alquier T, Furukawa N, Kim YB, Lee A, Xue B, Mu J, Foufelle F, Ferre P, Birnbaum MJ, Stuck BJ, Kahn BB: AMP-kinase regulates food intake by responding to hormonal and nutrient signals in the hypothalamus. Nature 2004;428:569.

53 Andersson U, Filipsson K, Abbott CR, Woods A, Smith K, Bloom SR, Carling D, Small CJ: AMP-activated protein kinase plays a role in the control of food intake. J Biol Chem 2004;279:12005.

54 Kola B, Hubina E, Tucci SA, Kirkham TC, Garcia EA, Mitchell SE, Williams LM, Hawley SA, Hardie DG, Grossman AB, Korbonits M: Cannabinoids and ghrelin have both central and peripheral metabolic and cardiac effects via AMP-activated protein kinase. J Biol Chem 2005;280:25196.

55 Kubota N, Yano W, Kubota T, Yamauchi T, Itoh S, Kumagai H, Kozono H, Takamoto I, Okamoto S, Shiuchi T, et al: Adiponectin stimulates AMP-activated protein kinase in the hypothalamus and increases food intake. Cell Metab 2007;6:55.

56 Steinberg GR, Watt MJ, Fam BC, Proietto J, Andrikopoulos S, Allen AM, Febbraio MA, Kemp BE: Ciliary neurotrophic factor suppresses hypothalamic AMP-kinase signaling in leptin-resistant obese mice. Endocrinology 2006;147:3906.

57 López M, Lelliott CJ, Tovar S, Kimber W, Gallego R, Virtue S, Blount M, Vázquez MJ, Finer N, Powles T, O'Rahilly S, Saha AK, Diéguez C, Vidal-Puig AJ: Tamoxifen-induced anorexia is associated with fatty acid synthase inhibition in the ventromedial nucleus of the hypothalamus and accumulation of malonyl-CoA. Diabetes 2006;55:1327.

58 Wolfgang MJ, Kurama T, Dai Y, Suwa A, Asaumi M, Matsumoto S, Cha SH, Shimokawa T, Lane MD: The brain-specific carnitine palmitoyltransferase-1c regulates energy homeostasis. Proc Natl Acad Sci USA 2006;103:7282.

59 Chakravarthy MV, Zhu Y, López M, Yin L, Wozniak DW, Coleman T, Hu Z, Wolfgang M, Vidal-Puig A, Lane MD, Semenkovich CF: Brain fatty acid synthase activates PPAR-alpha to maintain energy homeostasis. J Clin Invest 2007;117:2539.

60 Seo S, Ju S, Chung H, Lee D, Park S: Acute effects of glucagon-like peptide-1 on hypothalamic neuropeptide and AMP activated kinase expression in fasted rats. Endocr J 2008;55:867.

61 Shimizu H, Arima H, Watanabe M, Goto M, Banno R, Sato I, Ozaki N, Nagasaki H, Oiso Y: Glucocorticoids increase neuropeptide Y and agouti-related peptide gene expression via AMP-activated protein kinase signaling in the arcuate nucleus of rats. Endocrinology 2008;149:4544.

62 Kola B, Farkas I, Christ-Crain M, Wittmann G, Lolli F, Amin F, Harvey-White J, Liposits Z, Kunos G, Grossman AB, Fekete C, Korbonits M: The orexigenic effect of ghrelin is mediated through central activation of the endogenous cannabinoid system. PLoS ONE 2008;3:e1797.

63 Ishii S, Kamegai J, Tamura H, Shimizu T, Sugihara H, Oikawa S: Triiodothyronine (T3) stimulates food intake via enhanced hypothalamic AMP-activated kinase activity. Regul Pept 2008;151:164.

64 Vázquez MJ, González CR, Varela L, Lage R, Tovar S, Sangiao-Alvarellos S, Williams LM, Vidal-Puig A, Nogueiras R, López M, Diéguez C: Central resistin regulates hypothalamic and peripheral lipid metabolism in a nutritional-dependent fashion. Endocrinology 2008;149:4534.

65 López M, Saha AK, Diéguez C, Vidal-Puig A: The AMPK-malonyl-CoA-CPT1 axis in the control of hypothalamic neuronal function: reply. Cell Metab 2008;8:176.

66 Andrews ZB, Liu ZW, Walllingford N, Erion DM, Borok E, Friedman JM, Tschop MH, Shanabrough M, Cline G, Shulman GI, Coppola A, Gao XB, Horvath TL, Diano S: UCP2 mediates ghrelin's action on NPY/AgRP neurons by lowering free radicals. Nature 2008;454:846.

67 Claret M, Smith MA, Batterham RL, Selman C, Choudhury AI, Fryer LG, Clements M, Al Qassab H, Heffron H, Xu AW, Speakman JR, Barsh GS, Viollet B, Vaulont S, Ashford ML, Carling D, Withers DJ: AMPK is essential for energy homeostasis regulation and glucose sensing by POMC and AgRP neurons. J Clin Invest 2007;117:2325.

68 Anderson KA, Ribar TJ, Lin F, Noeldner PK, Green MF, Muehlbauer MJ, Witters LA, Kemp BE, Means AR: Hypothalamic CaMKK2 contributes to the regulation of energy balance. Cell Metab 2008;7: 377.

69 Tong Q, Ye C, McCrimmon RJ, Dhillon H, Choi B, Kramer MD, Yu J, Yang Z, Christiansen LM, Lee CE, Choi CS, Zigman JM, Shulman GI, Sherwin RS, Elmquist JK, Lowell BB: Synaptic glutamate release by ventromedial hypothalamic neurons is part of the neurocircuitry that prevents hypoglycemia. Cell Metab 2007;5:383.

Professor Carlos Diéguez, MD, PhD
Department of Physiology, School of Medicine, University of Santiago de Compostela
S. Francisco s/n
ES–15782 Santiago de Compostela (A Coruña) (Spain)
Tel. +34 981 582658, Fax +34 981 574145, E-Mail carlos.dieguez@usc.es

Ghrelin and Anterior Pituitary Function

Fabio Lanfranco · Giovanna Motta · Matteo Baldi ·
Valentina Gasco · Silvia Grottoli · Andrea Benso ·
Fabio Broglio · Ezio Ghigo

Division of Endocrinology, Diabetology and Metabolism, Department of Internal Medicine, University of Turin, Turin, Italy

Abstract

Ghrelin, a 28-amino-acid octanoylated peptide predominantly produced by the stomach, was discovered to be the natural ligand of the type 1a GH secretagogue receptor. Thus, it was considered as a natural GH secretagogue (GHS) additional to GHRH, although later on ghrelin has mostly been considered a major orexigenic factor. The GH-releasing action of ghrelin takes place both directly on pituitary cells and through modulation of GHRH from the hypothalamus; some functional anti-somatostatin action has also been shown. However, even at the neuroendocrine level, ghrelin is much more than a natural GHS. In fact, it significantly stimulates prolactin secretion in humans, independent of both gender and age and probably involving a direct action on somatomammotroph cells. Above all, ghrelin and synthetic GHS possess an acute stimulatory effect on the activity of the hypothalamus-pituitary-adrenal axis in humans, which is, at least, similar to that of the opioid antagonist naloxone, arginine vasopressin and even corticotropin-releasing hormone. Also, ghrelin plays a relevant role in the modulation of the hypothalamic-pituitary-gonadal function, with a predominantly CNS-mediated inhibitory effect upon the gonadotropin pulsatility both in animals and in humans.

Copyright © 2010 S. Karger AG, Basel

Ghrelin is a 28 amino acid peptide predominantly produced by the stomach but also expressed by several other tissues such as bowel, pancreas, kidney, immune system, placenta, testis, lung and hypothalamus [1–3]. Ghrelin is the first natural hormone to be identified in which the hydroxyl group of one of its serine-3 residue is acylated by n-octanoic acid [1]. This acylation is essential for hormone binding to the type 1a growth hormone secretagogue receptor (GHS-R1a), for the GH-releasing capacity of ghrelin, and most likely for its other endocrine actions [1] as well as for its orexigenic and metabolic actions [4, 5]. The nonacylated form ghrelin (UAG) is present in circulation in far greater amounts than its acylated form, does not bind GHS-R1a and is devoid of any neuroendocrine action. Nevertheless, UAG is an active peptide exerting metabolic as well as nonendocrine actions including cardiovascular and antiproliferative effects [4, 5]; as UAG does not bind GHS-R1a, these actions are likely mediated by a GHS-R subtype.

Apart from stimulating GH secretion, ghrelin and many synthetic GHS: (1) exhibit hypothalamic activities that result in stimulation of PRL and ACTH secretion; (2) negatively influence the pituitary-gonadal axis at both the central and the peripheral level; (3) stimulate appetite and a positive energy balance; (4) influence sleep and behavior; (5) control gastric motility and acid secretion, and (6) modulate pancreatic exocrine and endocrine function and affect glucose levels [4–6]. Clearly, ghrelin represents one of the best examples of a gastroenteropancreatic hormone playing a relevant role in the control of appetite, food intake and energy expenditure but meantime exerting a major role in glucose metabolism [4, 5]. This review, however, will specifically focus on the neuroendocrine actions of acylated ghrelin. Although ghrelin and GHS-R1a KO mice are not anorexic dwarfs [7], ghrelin plays some relevant physiological roles including at the neuroendocrine level.

Growth Hormone-Releasing Action

Ghrelin as well as synthetic GHS possess a strong and dose-related GH-releasing activity that is more marked in humans than in animals [1, 4–6, 8]. Natural and synthetic GHS stimulate GH release from somatotroph cells in vitro, probably by depolarizing the somatotroph membrane and by increasing the amount of GH secreted per cell [9, 10]. At the hypothalamic level, ghrelin and GHS act via mediation of GHRH-secreting neurons as indicated by evidence that passive immunization against GHRH, as well as pretreatment with GHRH antagonists, reduces their stimulatory effect on GH secretion [11–13]. Some functional anti-somatostatin action has also been shown [4, 5].

The GH-releasing effect of ghrelin and GHS undergoes marked age-related variations, increasing at puberty, persisting similar in adulthood and decreasing with aging; variations in estrogenic levels, GHRH hypoactivity and somatostatinergic hyperactivity would explain these age-related changes [5, 14, 15]. At variance with GHRH, the stimulatory effect of acylated ghrelin on GH secretion is reduced both in obese and in anorexic patients [5]. Acylated ghrelin has, any way, some physiological role in the generation of normal GH pulsatility [5, 16].

Acylated ghrelin as well as synthetic GHS could have diagnostic and therapeutic implications based on the strong and reproducible GH-releasing effects. Particularly when combined with GHRH, ghrelin and GHS could be used as potent and reliable provocative tests to evaluate the capacity of the pituitary to release GH for the diagnosis of GHD [4, 17]. Long-acting and orally active ghrelin analogs might represent an anabolic treatment in frail elderly subjects or in catabolic patients. At present, however, there is no definite evidence that shows the therapeutic efficacy of ghrelin analogs as GH/IGF-I axis-mediated anabolic agents in humans.

Prolactin- and Corticotrophin-Releasing Actions

Ghrelin likely modulates also lactotroph and corticotroph secretion in humans as well as in animals [4, 5].

Acylated ghrelin significantly stimulates PRL secretion in vitro from pituitary cell cultures probably acting on somatomammotroph cells [18]. Ghrelin significantly stimulates PRL secretion also in humans independently of both gender and age. The magnitude of the PRL-releasing action of ghrelin in humans is far lower than that of dopaminergic antagonists and TRH but similar to that of arginine [4, 5].

On the other hand, the stimulatory effect of ghrelin and synthetic GHS on the hypothalamus-pituitary-adrenal (HPA) axis in humans is remarkable and similar to that of the administration of naloxone, vasopressin, and even CRH [4, 5, 19]. The GHS-induced ACTH release is independent of gender but shows peculiar age-related variations, increasing at puberty, then showing a reduction in adulthood and, again, a trend toward an increase in aging [15, 20].

GHS do not stimulate ACTH release directly from pituitary cell cultures and their stimulatory effect on the HPA axis is lost after pituitary stalk section in pigs; thus, ghrelin stimulates the HPA axis via the CNS [4, 5]. In fact, ghrelin is likely to act at the hypothalamic level via stimulation of either CRH or arginine-vasopressin (AVP) [21, 22].

The ACTH response to natural and synthetic GHS is generally sensitive to the negative cortisol feedback mechanism [14, 20]. However, the stimulatory effect of ghrelin and GHS on corticotroph secretion is exaggerated and far higher than that of human CRH in patients with pituitary ACTH-dependent Cushing's disease [14, 20, 23, 24]. This unexpected stimulation of ACTH is peculiarly mediated by direct ghrelin action on neoplastic ACTH-secreting cells [25, 26].

Inhibitory Action of Ghrelin on Gonadotropin Secretion

Several in vitro and in vivo animal studies suggest that the ghrelin system negatively influences the gonadal axis [27–34]. In fact, acylated ghrelin suppresses LH pulsatility in rodent, ovine and primate models [27, 29–34]. Ghrelin has been shown to decrease LH responsiveness to GnRH from the pituitary in vitro [33]. However, ghrelin infusion decreased LH pulse frequency but not pulse amplitude in adult ovariectomized rhesus primates, suggesting that ghrelin could inhibit the GnRH pulse activity [32].

We showed that a prolonged infusion of acylated ghrelin quantitatively and qualitatively inhibited LH but not FSH secretion in healthy young males; in fact, the infusion of the peptide was associated with clear inhibition of LH mean concentration and pulsatility [35]. In contrast with in vitro data showing that ghrelin reduces the LH response to GnRH in rodents [33], in humans the LH response to this neurohormone is not modified by the exposure to acylated ghrelin; these findings are therefore against the hypothesis that ghrelin plays any direct inhibitory role on pituitary gonadotropic cells. As acylated ghrelin inhibits the gonadotropin response to naloxone in humans, this clearly points toward a CNS-mediated inhibitory action on the human gonadal axis [35].

The inhibitory effect of a gastroenteropancreatic hormone, like ghrelin, on the gonadal axis fits well with clinical data in pathophysiological conditions. Anorexia nervosa, malnutrition and cachexia are generally associated to hypogonadism that reflects a functional impairment of neuroendocrine mechanisms [36]. Metabolic factors have a major impact on the regulation of ghrelin secretion, and the pathophysiological conditions mentioned above are not associated with ghrelin hypersecretion by chance [37, 38]. Thus, it seems reasonable to hypothesize that ghrelin hypersecretion could have a role in the functional hypogonadism connoting anorexia, malnutrition and cachexia.

References

1 Kojima M, Hosoda H, Date Y, Nakazato M, Matsuo H, Kangawa K: Ghrelin is a growth-hormone acylated peptide from stomach. Nature 1999;402:656–660.
2 Kojima M, Hosoda H, Matsuo H, Kangawa K: Ghrelin: discovery of the natural endogenous ligand for the growth hormone secretagogue receptor. Trends Endocrinol Metab 2001;12:118–122.
3 Kojima M, Kangawa K: Ghrelin: structure and function. Physiol Rev 2005;85:495–522.
4 van der Lely AJ, Tschöp M, Heiman ML, Ghigo E: Biological, physiological, pathophysiological, and pharmacological aspects of ghrelin. Endocr Rev 2004;25:426–457.
5 Ghigo E, Broglio F, Arvat E, Maccario M, Papotti M, Muccioli G: Ghrelin: more than a natural GH secretagogue and/or an orexigenic factor. Clin Endocrinol 2005;62:1–17.
6 Broglio F, Prodam F, Riganti F, Muccioli G, Ghigo E: Ghrelin: from somatotrope secretion to new perspectives in the regulation of peripheral metabolic functions. Front Horm Res 2006;35:102–114.
7 Sun Y, Ahmed S, Smith RG: Deletion of ghrelin impairs neither growth nor appetite. Mol Cell Biol 2003;23:7973–7981.
8 Arvat E, Di Vito L, Broglio F, Papotti M, Muccioli G, Dieguez C, Casanueva FF, Deghenghi R, Camanni F, Ghigo E: Preliminary evidence that ghrelin, the natural GH secretagogue (GHS)-receptor ligand, strongly stimulates GH secretion in humans. J Endocrinol Invest 2000;23:493–495.
9 Goth MI, Lyons CE, Canny BJ, Thorner MO: Pituitary adenylate cyclase activating polypeptide, growth hormone (GH)-releasing peptide and GH-releasing hormone stimulate GH release through distinct pituitary receptors. Endocrinology 1992;130:939–944.
10 Smith RG, Cheng K, Schoen WR, Pong SS, Hickey G, Jacks T, Butler B, Chan WW, Chaung LY, Judith F: A nonpeptidyl growth hormone secretagogue. Science 1993;260:1640–1643.
11 Clark RG, Carlsson MS, Trojnar J, Robinson IC: The effects of a growth hormone-releasing peptide and growth hormone-releasing factor in conscious and anaesthetized rats. J Neuroendocrinol 1989;1:249–255.
12 Bowers CY, Sartor AO, Reynolds GA, Badger TM: On the actions of the growth hormone-releasing hexapeptide, GHRP. Endocrinology 1991;128:2027–2035.
13 Barkan AL, Dimaraki EV, Jessup SK, Symons KV, Ermolenko M, Jaffe CA: Ghrelin secretion in humans is sexually dimorphic, suppressed by somatostatin, and not affected by the ambient growth hormone levels. J Clin Endocrinol Metab 2003;88:2180–2184.
14 Ghigo E, Arvat E, Giordano R, Broglio F, Gianotti L, Maccario M, Bisi G, Graziani A, Papotti M, Muccioli G, Deghenghi R, Camanni F: Biologic activities of growth hormone secretagogues in humans. Endocrine 2001;14:87–93.
15 Broglio F, Benso A, Castiglioni C, Gottero C, Prodam F, Destefanis S, Gauna C, Van Der Lely AJ, Deghenghi R, Bo M, Arvat E, Ghigo E: The endocrine response to ghrelin as a function of gender in humans in young and elderly subjects. J Clin Endocrinol Metab 2003;88:1537–1542.

16 Zizzari P, Halem H, Taylor J, Dong JZ, Datta R, Culler MD, Epelbaum J, Bluet-Pajot MT: Endogenous ghrelin regulates episodic growth hormone (GH) secretion by amplifying GH pulse amplitude: evidence from antagonism of the GH secretagogue-R1a receptor. Endocrinology 2005; 146:3836–3842.

17 Ghigo E, Arvat E, Aimaretti G, Broglio F, Giordano R, Camanni F: Diagnostic and therapeutic uses of growth hormone-releasing substances in adult and elderly subjects. Baillieres Clin Endocrinol Metab 1998;12:341–358.

18 Bowers CY: On a peptidomimetic growth hormone-releasing peptide. J Clin Endocrinol Metab 1994; 79:940–942.

19 Arvat E, Maccario M, Di Vito L, Broglio F, Benso A, Gottero C, Papotti M, Muccioli G, Dieguez C, Casanueva FF, Deghenghi R, Camanni F, Ghigo E: Endocrine activities of ghrelin, a natural growth hormone secretagogue (GHS), in humans: comparison and interactions with hexarelin, a nonnatural peptidyl GHS, and GH-releasing hormone. J Clin Endocrinol Metab 2001;86:1169–1174.

20 Arvat E, Ramunni J, Bellone J, Di Vito L, Baffoni C, Broglio F, Deghenghi R, Bartolotta E, Ghigo E: The GH, prolactin, ACTH and cortisol responses to hexarelin, a synthetic hexapeptide, undergo different age-related variations. Eur J Endocrinol 1997; 137:635–642.

21 Korbonits M, Little JA, Forsling ML, Tringali G, Costa A, Navarra P, Trainer PJ, Grossman AB: The effect of growth hormone secretagogues and neuropeptide Y on hypothalamic hormone release from acute rat hypothalamic explants. J Neuroendocrinol 1999;11:521–528.

22 Mozid AM, Tringali G, Forsling ML, Hendricks MS, Ajodha S, Edwards R, Navarra P, Grossman AB, Korbonits M: Ghrelin is released from rat hypothalamic explants and stimulates corticotrophin-releasing hormone and arginine-vasopressin. Horm Metab Res 2003;35:455–459.

23 Arvat E, Giordano R, Ramunni J, Arnaldi G, Colao A, Deghenghi R, Lombardi G, Mantero F, Camanni F, Ghigo E: Adrenocorticotropin and cortisol hyperresponsiveness to hexarelin in patients with Cushing's disease bearing a pituitary microadenoma, but not in those with macroadenoma. J Clin Endocrinol Metab 1998;83:4207–4211.

24 Leal-Cerro A, Torres E, Soto A, Dios E, Deghenghi R, Arvat E, Ghigo E, Dieguez C, Casanueva FF: Ghrelin is no longer able to stimulate growth hormone secretion in patients with Cushing's syndrome but instead induces exaggerated corticotropin and cortisol responses. Neuroendocrinology 2002;76: 390–396.

25 Barlier A, Zamora AJ, Grino M, Gunz G, Pellegrini-Bouiller I, Morange-Ramos I, Figarella-Branger D, Dufour H, Jaquet P, Enjalbert A: Expression of functional growth hormone secretagogue receptors in human pituitary adenomas: polymerase chain reaction, triple in-situ hybridization and cell culture studies. J Neuroendocrinol 1999;11:491–502.

26 Martínez-Fuentes AJ, Moreno-Fernández J, Vázquez-Martínez R, Durán-Prado M, de la Riva A, Tena-Sempere M, Diéguez C, Jiménez-Reina L, Webb SM, Pumar A, Leal-Cerro A, Benito-López P, Malagón MM, Castaño JP: Ghrelin is produced by and directly activates corticotrope cells from adrenocorticotropin-secreting adenomas. J Clin Endocrinol Metab 2006;91:2225–2231.

27 Furuta M, Funabashi T, Rimura F: Intracerebroventricular administration of ghrelin rapidly suppresses pulsatile luteinizing hormone secretion in ovariectomized rats. Biochem Biophys Res Commun 2001;288:780–785.

28 Tena-Sempere M, Barreiro ML, Gonzalez LC, Gaytan F, Zhang FP, Caminos JE, Pinilla L, Casanueva FF, Dieguez C, Aguilar E: Novel expression and functional role of ghrelin in rat testis. Endocrinology 2002;143:717–725.

29 Kawamura K, Sato N, Fukuda J, Kodama H, Kumegai J, Tanikawa H, Nakamura A, Honda Y, Sato T, Tanaka T: Ghrelin inhibits the development of mouse preimplantation embryos in vitro. Endocrinology 2003;144:2623–2633.

30 Barreiro ML, Tena-Sempere M: Ghrelin and reproduction: a novel signal linking energy status and fertility? Mol Cell Endocrinol 2004;226:1–9.

31 Fernández-Fernández R, Tena-Sempere M, Aguilar E, Pinilla L: Ghrelin effects on gonadotropin secretion in male and female rats. Neurosci Lett 2004;362: 103–107.

32 Vulliémoz NR, Xiao E, Xia-Zhang L, Germond M, Rivier J, Ferin M: Decrease in luteinizing hormone pulse frequency during a five-hour peripheral ghrelin infusion in the ovariectomized rhesus monkey. J Clin Endocrinol Metab 2004;89:5718–5723.

33 Fernández-Fernández R, Tena-Sempere M, Navarro VM, Barreiro ML, Castellano JM, Aguilar E, Pinilla L: Effects of ghrelin upon gonadotropin-releasing hormone and gonadotropin secretion in adult female rats: in vivo and in vitro studies. Neuroendocrinology 2005;82:245–255.

34 Iqbal J, Kurose Y, Canny B, Clarke IJ: Effects of central infusion of ghrelin on food intake and plasma levels of growth hormone, luteinizing hormone, prolactin, and cortisol secretion in sheep. Endocrinology 2006;147:510–519.

35 Lanfranco F, Bonelli L, Baldi M, Me E, Broglio F, Ghigo E: Acylated ghrelin inhibits spontaneous LH pulsatility and responsiveness to naloxone, but not that to GnRH in young men: evidence for a central inhibitory action of ghrelin on the gonadal axis. J Clin Endocrinol Metab 2008;93:3633–3639.

36 Vanhorebeek I, Langouche L, Van den Berghe G: Endocrine aspects of acute and prolonged critical illness. Nat Clin Pract Endocrinol Metab 2006;2:20–31.

37 Broglio F, Gianotti L, Destefanis S, Fassino S, Abbate Daga G, Mondelli V, Lanfranco F, Gottero C, Gauna C, Hofland L, Van der Lely AJ, Ghigo E: The endocrine response to acute ghrelin administration is blunted in patients with anorexia nervosa, a ghrelin hypersecretory state. Clin Endocrinol 2004;60:592–599.

38 Shimizu Y, Nagaya N, Isobe T, Imazu M, Okumura H, Hosoda H, Kojima M, Kangawa K, Kohno N: Increased plasma ghrelin level in lung cancer cachexia. Clin Cancer Res 2003;9:774–778.

Ezio Ghigo, MD
Division of Endocrinology and Metabolism, Department of Internal Medicine
University of Turin, Corso Dogliotti 14
IT–10126 Torino (Italy)
Tel. +39 0 11 6334317, Fax +39 0 11 6647421, E-Mail ezio.ghigo@unito.it

Author Index

Alexandraki, K.I. 165
Alvarez, C.V. 127
Arzt, E. IX, 1, 158
Auriemma, R.S. 94

Baldi, M. 206
Ballarino, C. 42
Balzaretti, M. 50
Barahona, M.J. 152
Basavilbaso, N.G. 145
Basso, A. 42, 145
Becu-Villalobos, D. 59
Ben-Shlomo, A. 7
Benso, A. 206
Bilodeau, S. 15
Bravo, S.B. 127
Broglio, F. 206
Bronstein, M.D. IX, 70, 174
Bruera, D. 42
Bruno, O. 42
Buchfelder, M. 109

Carabelli, A. 42
Carrizo, A. 50
Casanueva, F.F. 139
Castro, A.I. 139
Cescato, V.A.S. 70
Chervin, A. 42
Chesnokova, V. 7
Chiovato, L. 139
Christiansen, S. 50
Clapp, C. 184
Colao, A. 94

da Boit, K. 196
Danilowitz, K. 42
de Oliveira, E. 70

Diaz-Rodriguez, E. 127
Diaz-Torga, G. 59
Diéguez, C. 196
Drouin, J. 15

Faggiano, A. 94
Fainstein Day, P. 50
Fidalgo, S. 42
Fideleff, H. 42
Frohman, L.A. 121
Fuertes, M. 1

Gabellieri, E. 139
Gadelha, M.R. 87, 121
García Basavilbaso, N. 42
Garcia-Lavandeira, M. 127
Garcia-Rendueles, M.E.R. 127
García-Tornadu, I. 59
Gasco, V. 206
Gerez, J. 1
Ghigo, E. 206
Giacomini, D. 1
Glerean, M. 50
Grossman, A.B. VII, 165
Grottoli, S. 206
Guelman, R. 42
Guitelman, M. IX, 42, 145

Haedo, M. 1, 158

Jaita, G. 25

Katz, D. 42
Kozak, A. 50

Labeur, M. 1, 158
Lage, M. 139

Lanfranco, F. 206
Leal, R. 42
Librandi, F. 42
Lima, G.A.B. 87
Lombardi, G. 94
López, M. 196
Lourenço, Jr., D.M. 77
Lovazzano, S. 50
Low, M.J. 59

Magri, L. 25
Mallea Gil, S. 42
Manavela, M. 42
Martínez de la Escalera, G. 184
Martínez de Morentin, P.B. 196
Melmed, S. 7
Méndez, I. 184
Moreno-Carranza, B. 184
Motta, G. 206
Musolino, N.R.C. 70

Nagelberg, A. 42
Noain, D. 59
Nogueiras, R. 196
Novelle, M.G. 196

Ochoa, A.S. 94
Oneto, A. 145

Paez Pereda, M. 158
Páez-Pereda, M. 1
Perez-Millan, M.I. 59
Perez-Romero, S. 127
Perone, M.J. 158
Pietrani, M. 50
Pinto, E.M. 70
Pisera, D. 25
Pivonello, R. 94

Radczuk, G. 145
Radl, D. 25
Renner, U. 158
Resmini, E. 152
Risso, G. 59
Rodrigues, J.S. 127
Roussel-Gervais, A. 15
Rubinstein, M. 59

Salgado, L.R. 70
Sarkar, D.K. 32
Schlaffer, S.-M. 109
Seilicovich, A. 25
Servidio, M. 42
Soares, I.C. 70
Stalla, G.K. 1, 158
Stalldecker, G. 42
Strasburger, C.J. 87
Sucunza, N. 152

Theodoropoulou, M. 158
Toledo, R.A. 77
Toledo, S.P.A. 77

van der Lelij, A.J. 190
Vega, C. 184
Vitale, M. 42

Wakamatsu, A. 70
Wawrowsky, K. 7
Webb, S.M. 152
Wu, Z. 87

Zaldivar, V. 25
Zamorano, M. 184
Zárate, S. 25
Zonis, S. 7

Subject Index

Acromegaly
   epidemiology 174
   familial, see Isolated familial
      somatotropinoma
   growth hormone isoforms 91
   insulin-like growth factor-I monitoring
      clinical application 147–150
      overview 145, 146
      serum assays 146–149
   morbidity and mortality 174, 175
   treatment
      combination therapies
         somatostatin analogs/
            cabergoline 178, 179
         somatostatin analogs/
            pegvisomant 177, 178
         somatostatin analogs/surgery
            179–181
      dopamine agonists 61, 175, 176
      goals 175
      pegvisomant therapy
         adverse events 47, 48
         insulin-like growth factor-I
            response 46
         long-term studies 43
         mechanism of action 176, 177
         study design 44–46
      pharmacotherapy overview 42, 43
      radiotherapy 175
      somatostatin analogs 176
      surgery 175
ACTH, see Adrenocorticotropic hormone
Adrenocorticotropic hormone (ACTH),
   ghrelin mediation of release 208
Agouti-related peptide (AgRP), ghrelin
   mediation of food intake 197, 198
AgRP, see Agouti-related peptide

AIP, see Aryl hydrocarbon receptor-interacting
   protein
Alcohol abuse, hyperprolactinemia
   association
   effects
      men 33
      primates 33
      women 33
   lactotrope mechanisms
      dopamine D2 receptor splice
         variants 34–36
      fibroblast growth factor-2 from
         folliculo-stellate cells 38, 39
      G protein activation 36, 37
      transforming growth factor-β 37, 38
Anterior pituitary
   cell renewal 25, 26
   estrogen
      mechanisms of cell turnover 26, 27
      rapid actions on cells 29–31
Apoptosis-stimulating proteins of p53 (ASPP)
   family members 72
   inhibitor, see IASPP
Aryl hydrocarbon receptor-interacting
   protein (AIP)
   familial isolated pituitary adenoma
      mutations 79–81, 84
   familial somatotropinoma mutations
      122–124
ASPP, see Apoptosis-stimulating proteins
   of p53

BIM-23A387, pituitary tumor treatment 160
BIM-23A760, pituitary tumor treatment
   160, 161
Bromocriptine, see Dopamine
Bsx, expression regulation by ghrelin 199

Cabergoline, see Dopamine
Carcinoma, see Pituitary carcinoma
Carney complex (CNC)
    genetics 77–79
    growth hormone-secreting tumors 121
Cephalic phase insulin release (CPIR) 192, 193
Ciliary neurotrophic factor (CNTF), pituitary receptors 2
CNC, see Carney complex
CNTF, see Ciliary neurotrophic factor
Computed tomography (CT), pituitary adenoma 118
CPIR, see Cephalic phase insulin release
CS, see Cushing's syndrome
CT, see Computed tomography
Cushing's syndrome (CS)
    health-related quality of life 155, 156
    morbidity and mortality
        prospective case-control study 153–155
        retrospective study 153
    prognosis 152, 153
    treatment
        cortisol inhibitors 166–168
        dopamine agonists 168, 169
        interferon-γ 162
        mifepristone 168
        overview 165, 166
        retinoic acid 161
        somatostatin analogs 169, 170

Dependence receptors, see RET receptor
Dietary restriction (DR), life span effects 192
Differential display, see Messenger RNA differential display
Dopamine
    agonists
        acromegaly treatment 61, 175, 176
        Cushing's syndrome management 168, 169
        non-lactotroph pituitary tumor treatment 159
        pituitary carcinoma management 102
        growth hormone regulation 60, 61, 66, 67
    receptors
        brain distribution 60
        D2 receptor
            knockout mouse and growth phenotype 61–66
            splice variants in hyperprolactinemia from alcohol abuse 34–36
        types 60
DR, see Dietary restriction

EGFR, see Epidermal growth factor receptor
Epidermal growth factor receptor (EGFR), pituitary carcinoma proliferation marker 98, 99
Estrogen
    anterior pituitary effects
        mechanisms of cell turnover 26, 27
        rapid actions on cells 29–31
        receptor splice variants in pituitary 28
Etomidate, Cushing's syndrome management 167

Familial isolated pituitary adenoma (FIPA), see also Isolated familial somatotropinoma
    aryl hydrocarbon receptor-interacting protein mutations 79–81, 84
    definition 77
    genetics 78, 79
    modifier genes 83
    susceptibility loci 81–83
FGF-2, see Fibroblast growth factor-2
Fibroblast growth factor-2 (FGF-2), hyperprolactinemia from alcohol abuse role 38, 39
FIPA, see Familial isolated pituitary adenoma
Fluconazole, Cushing's syndrome management 167
Follicle-stimulating hormone (FSH), ghrelin inhibition of secretion 208, 209
Folliculo-stellate cell, hyperprolactinemia from alcohol abuse role 38, 39
FSH, see Follicle-stimulating hormone

G protein
    activation in hyperprolactinemia from alcohol abuse 36, 37
    oncogene, see GSP oncogene
Galectin-3 99
GFRa1, see RET receptor
GH, see Growth hormone
Ghrelin
    Bsx expression regulation 199
    food intake and body weight homeostasis regulation 196, 197
    forms 206, 207
    gonadotropin secretion inhibition 208
    hormone releasing activity
        adrenocorticotropic hormone 208
        growth hormone 207
        prolactin 208

Ghrelin (continued)
  hypothalamic lipid metabolism
    effects 199–201
  modulators in food intake regulation
    agouti-related peptide 197, 198
    neuropeptide Y 197, 198
  receptor in hypothalamus 197
  tissue distribution 206
GHRH, see Growth hormone-releasing hormone
Glial-derived neurotrophic factor, see RET receptor
GLP-1, see Glucagon-like peptide-1
Glucagon-like peptide-1 (GLP-1), taste bud signaling 192
gp130, pituitary tumorigenesis role 2
Growth hormone (GH)
  antagonist, see Pegvisomant
  deficiency
    aging 139, 140
    morbidity 139
    testing in adults 140–143
  dopamine regulation 60, 61, 66, 67
  excess, see Acromegaly
  ghrelin mediation of release 207
  isoforms
    acromegaly 91
    assays
      20K-hGH 90
      22K-hGH 90
      non-22K-hGH 89
    20K-hGH structure and function 88, 89
    overview 87, 88
Growth hormone-releasing hormone (GHRH), dopamine regulation of action 64–66
GSP oncogene, pituitary carcinoma role 97

Hyperprolactinemia
  alcohol abuse association
    effects
      men 33
      primates 33
      women 33
    lactotrope mechanisms
      dopamine D2 receptor splice variants 34–36
      fibroblast growth factor-2 from folliculo-stellate cells 38, 39
      G protein activation 36, 37
      transforming growth factor-β 37, 38

  clinical manifestations 32
  definition 32

IASPP
  functional overview 72
  pituitary tumor expression 73–75
IFN-α, see Interferon-α
IFN-γ, see Interferon-γ
IFS, see Isolated familial somatotropinoma
IGF-I, see Insulin-like growth factor-I
IL-6, see Interleukin-6
IL-11, see Interleukin-11
Insulin, cephalic phase insulin release 192, 193
Insulin-like growth factor-I (IGF-I)
  acromegaly monitoring
    clinical application 147–150
    overview 145, 146
    serum assays 146–149
  pegvisomant response 46
  physiological factors affecting levels 146
Insulin tolerance test (ITT), growth hormone deficiency testing 141
Interferon-α (IFN-α), pituitary carcinoma management 103
Interferon-γ (IFN-γ), Cushing's syndrome management 162
Interleukin-6 (IL-6)
  family of cytokines 2
  gp130 receptor 2
  pituitary adenoma role 3, 12
Interleukin-11 (IL-11), pituitary receptors 2
Isolated familial somatotropinoma (IFS)
  aryl hydrocarbon receptor-interacting protein mutations 122–124
  clinical features 124, 125
ITT, see Insulin tolerance test

Ketoconazole, Cushing's syndrome management 166, 167
Ki-67
  pituitary carcinoma proliferation marker 98
  prolactinoma expression 56

Lanreotide
  acromegaly management 176
  pituitary carcinoma management 102
Leptin, taste interactions 193
Leukemia inhibitory factor (LIF), pituitary receptors 2
LH, see Luteinizing hormone

LIF, *see* Leukemia inhibitory factor
Luteinizing hormone (LH), ghrelin inhibition of secretion 208, 209

Magnetic resonance imaging (MRI)
    differential diagnosis of pituitary masses 116, 117
    intraoperative pituitary imaging 116
    overview 109, 110
    pituitary adenoma 110–114
    postoperative pituitary imaging 113–116
Messenger RNA differential display, pituitary gene cloning 3, 4
Metyrapone, Cushing's syndrome management 166
Mifepristone, Cushing's syndrome management 168
Mitotane, Cushing's syndrome management 167, 168
MRI, *see* Magnetic resonance imaging
Multiple endocrine neoplasia type 1
    genetics 77–79
    growth hormone-secreting tumors 121

Neuropeptide Y (NPY)
    ghrelin mediation of food intake 197, 198
    taste bud expression 194
NF-κB, *see* Nuclear factor-κB
nm23, pituitary carcinoma mutation 97
NPY, *see* Neuropeptide Y
Nuclear factor-κB (NF-κB)
    IASPP effects 75, 76
    inhibitors 73

Octreotide
    acromegaly management 176
    Cushing's syndrome management 169
    pituitary carcinoma management 102
Olfaction, metabolic interactions 191, 192

p27, pituitary carcinoma mutation 97
p53
    apoptosis-stimulating proteins of p53 family members 72
    binding proteins 71
    inhibitor, *see* IASPP
    pituitary carcinoma mutation 97, 99
p57Kip2, pituitary development role 19–22
Pasireotide
    Cushing's syndrome management 169
    pituitary tumor treatment 159, 160

PCNA, *see* Proliferating cell nuclear antigen
Pegvisomant
    acromegaly management
        adverse events 47, 48
        insulin-like growth factor-I response 46
        long-term studies 43
        mechanism of action 176, 177
        study design 44–46
    structure 43
PET, *see* Positron emission tomography
PIT, *see* RET receptor
Pituitary carcinoma
    clinical findings 100, 101
    diagnosis 99, 100
    epidemiology 95, 96
    etiology 96
    pathogenesis
        GSP oncogene 97
        tumor suppressor genes 97
    proliferation markers
        adhesion molecules 98
        DNA ploidy status 99
        epidermal growth factor receptor 98, 99
        galectin-3 99
        Ki-67 98
        proliferating cell nuclear antigen 98
    treatment
        chemotherapy 103, 104
        pharmacotherapy 102, 103
        radiotherapy 101, 102
        surgery 101
Pituitary stem cells
    cell cycle
        exit in early development 18–21
        independent control of differentiation and cell cycle 22
        re-entry blockade in differentiated cells 21
    differentiation 16–18
    markers 15, 16
Pituitary tumor transforming gene (PTTG)
    cancer roles 10, 12
    cell cycle regulation 9
    knockout studies 10
Positron emission tomography (PET), pituitary adenoma 118, 119
PRL, *see* Prolactin
Prolactin (PRL)
    excess, *see* Hyperprolactinemia
    ghrelin mediation of release 208
    tumor secretion, *see* Prolactinoma

Prolactin (PRL) (continued)
  vasoinhibin fragments
    anterior pituitary 185, 186
    overview 184
    physiological actions 185
    posterior pituitary 186, 187
    processing 185
Prolactinoma, sex differences 50–57
Proliferating cell nuclear antigen (PCNA), pituitary carcinoma proliferation marker 98
PTTG, see Pituitary tumor transforming gene

Rb, see Retinoblastoma protein
RET receptor
  apoptosis induction through PIT 131–133
  dependence receptors 130, 131
  GFRa1 coreceptor 126
  glial-derived neurotrophic factor ligand 128
  knockout mouse 133, 134
  mutation in disease 128
  pituitary function 128–130
  prospects for study 134–136
  structure 127
Retinoblastoma protein (Rb), pituitary carcinoma mutation 97
Retinoic acid, Cushing's syndrome management 161
RSUME
  pituitary tumorigenesis role 4, 5
  posttranslational modification 4
  splice variants 4
  tissue distribution 4

Senescence
  mechanisms 8, 9, 12, 13
  overview 8
  p21 immunohistochemistry 10, 11
  pituitary tumor transforming gene knockout studies 10
Single photon emission computed tomography (SPECT), pituitary adenoma 118, 119
Sleep, disorders and metabolic/endocrine disturbances 190, 191
Smell, see Olfaction
Somatostatin analogs, see specific analogs
SPECT, see Single photon emission computed tomography
Stem cells, see Pituitary stem cells

Taste, metabolic interactions 192–194
Temozolamide, pituitary carcinoma management 103
TGF-β, see Transforming growth factor-β
Transforming growth factor-β (TGF-β), hyperprolactinemia from alcohol abuse role 37, 38

Vasoinhibins, see Prolactin